Student's Solutions M

to accompany

Elementary Number Theory

Sixth Edition

David M. Burton
University of New Hampshire

 **Higher Education**

Boston   Burr Ridge, IL   Dubuque, IA   Madison, WI   New York   San Francisco   St. Louis
Bangkok   Bogotá   Caracas   Kuala Lumpur   Lisbon   London   Madrid   Mexico City
Milan   Montreal   New Delhi   Santiago   Seoul   Singapore   Sydney   Taipei   Toronto

The McGraw·Hill Companies

Student's Solutions Manual to accompany
ELEMENTARY NUMBER THEORY, SIXTH EDITION
DAVID M. BURTON

Published by McGraw-Hill Higher Education, an imprint of The McGraw-Hill Companies, Inc., 1221 Avenue of the Americas,
New York, NY 10020. Copyright © 2007 by The McGraw-Hill Companies, Inc. All rights reserved.

No part of this publication may be reproduced or distributed in any form or by any means, or stored in a database or retrieval system,
without the prior written consent of The McGraw-Hill Companies, Inc., including, but not limited to, network or other electronic
storage or transmission, or broadcast for distance learning.

This book is printed on acid-free paper.

2 3 4 5 6 7 8 9 0 QPD/QPD 0 9 8 7 6

ISBN-13: 978-0-07-321962-2
ISBN-10: 0-07-321962-2

www.mhhe.com

Contents

Preface

This *Student's Solutions Manual* is intended to accompany the sixth edition of David M. Burton's *Elementary Number Theory*. It contains complete solutions to almost all of the odd-numbered problems in the text. The exceptions are certain numerical exercises whose answers are given in the back of the book, and a few others whose accompanying hints are sufficiently detailed as to provide the solutions. We have tried to be accurate; but there are some 380 odd-numbered problems, many with several parts, so that occasional errors and misprints would appear to be inevitable. Certain problems may also have shorter or more elegant solutions than the ones presented here. If you discover any way to improve a solution, we would appreciate hearing of it.

D.M.B.

Chapter 1

Some Preliminary Considerations

1.1 Mathematical Induction

1. (a) If $n = 1$, then $1 = \frac{1(1+1)}{2}$, so the assertion holds when n $= 1$. Suppose that for some k,

$$1 + 2 + \cdots + k = \frac{k(k+1)}{2}.$$

Then

$$
\begin{aligned}
1 + 2 + \cdots + k &= \frac{k(k+1)}{2} + (k+1) \\
&= (k+1)(\frac{k}{2} + 1) \\
&= \frac{(k+1)(k+1)}{2}.
\end{aligned}
$$

Hence the assertion is true for $k + 1$. This completes the induction step and proves the result for all positive integers n.

(b) The assertion is true for $n = 1$ since $1 = 1^2$. Suppose that the assertion holds for some k, so that

$$1 + 3 + \cdots + (2k - 1) = k^2.$$

Then

$$1 + 3 + \cdots + (2k - 1) + (2k + 1) = k^2 + (2k + 1) = (k + 1)^2.$$

1

Thus the statement is true for $k+1$, which completes the induction.

(c) If $n = 1$, then $1 \cdot 2 = 1 \cdot (1+1)(1+2)/3$, so the assertion is correct in this case. Suppose that for some k, we have

$$1 \cdot 2 + 2 \cdot 3 + \cdots + k(k+1) = \frac{k(k+1)(k+2)}{3};$$

then

$$
\begin{aligned}
1 \cdot 2 + 2 \cdot 3 + \quad \cdots \quad &+ k(k+1) + (k+1)(k+2) \\
&= \frac{k(k+1)(k+2)}{3} + (k+1)(k+2) \\
&= (k+1)(k+2)(\frac{k}{3}+1) \\
&= \frac{(k+1)(k+2)(k+3)}{3}.
\end{aligned}
$$

This shows that if the assertion is correct for k, then it is correct for $k+1$, thereby completing the induction.

(d) If $n = 1$, then $1^2 = 1 \cdot (4-1)/3$, so the assertion is true for $n = 1$. Assume that for some k,

$$1^2 + 3^2 + \cdots + (2k-1)^2 = k(4k^2 - 1)/3.$$

Then

$$
\begin{aligned}
1^2 + 3^2 + \quad \cdots \quad &+ (2k-1)^2 + (2k+1)^2 \\
&= \frac{k(4k^2 - 1)}{3} + (2k+1)^2 \\
&= (2k+1)\left[\frac{k(2k-1)}{3} + (2k+1)\right] \\
&= (2k+1)\left[\frac{(2k+3)(k+1)}{3}\right] \\
&= \frac{(k+1)(4(k+1)^2 - 1)}{3}.
\end{aligned}
$$

Thus the truth of the assertion for k implies its truth for $k+1$. This completes the induction.

(e) The assertion is true for $n = 1$, since

$$1^3 = \left[\frac{1(1+1)}{2}\right]^2.$$

If we know that

$$1^3 + 2^3 + \cdots + k^3 = \left[\frac{k(k+1)}{2}\right]^2$$

for a certain integer k, then

$$
\begin{aligned}
1^3 + 2^3 + \ \cdots \ + k^3 + (k+1)^3 &= \left[\frac{k(k+1)}{2}\right]^2 + (k+1)^3 \\
&= (k+1)^2 \left[\frac{k^2}{4} + (k+1)\right] \\
&= \frac{(k+1)^2(k+2)^2}{4}.
\end{aligned}
$$

This completes the induction and proves the assertion.

3. When $n = 1$, the assertion reduces to $a - 1 = a - 1$, hence is true in this case. Assume that the assertion holds for all positive integers less than or equal to k. Then

$$
\begin{aligned}
a^{k+1} - 1 &= (a+1)(a^k - 1) - a(a^k - 1) \\
&= (a+1)(a-1)(a^{k-1} + \cdots + a + 1) - a(a-1)(a^{k-2} + \cdots + a + 1) \\
&= (a-1)[(a+1)(a^{k-1} + \cdots + a + 1) - a(a^{k-2} + \cdots + a + 1)] \\
&= (a-1)(a^k + a^{k-1} + \cdots + a + 1),
\end{aligned}
$$

so the assertion holds for $k + 1$. This completes the induction.

7. Proceeding by induction, note that the assertion holds when $n = 1$; indeed, $1 \cdot 1! = (1+1)! - 1$. Suppose the formula holds for a given positive integer k; that is, $1 \cdot 1! + 2 \cdot 2! + \cdots + k \cdot k! = (k+1)! - 1$. Then

$$
\begin{aligned}
1 \cdot 1! \ + \ 2 \cdot 2! + \cdots + k \cdot k! + (k+1)(k+1)! \\
= \ (k+1)! - 1 + (k+1)(k+1)! \\
= \ [1 + (k+1)](k+1)! - 1 \\
= \ (k+2)! - 1.
\end{aligned}
$$

This shows that the formula holds for $k + 1$ whenever it holds for k, thereby completing the induction.

9. Again use induction, observing that the result is trivial when $n = 1$.

Suppose for some k that $(1+a)^k \geq 1 + ka$. Then

$$\begin{aligned}
(1+a)^{k+1} &= (1+a)^k(1+a) \geq (1+ka)(1+a) \\
&\geq 1 + (k+1)a + a^2 \geq 1 + (k+1)a.
\end{aligned}$$

Thus the inequality holds for $k+1$ whenever it holds for k.

11. Use an induction argument to show that $(2n)!/2^n n!$ is an integer for $n \geq 0$. Since $0! = 1$, the assertion is clearly true for $n = 0$. Assume that it holds for some k and consider the case $k+1$. Then

$$\frac{(2(k+1))!}{2^{k+1}(k+1)!} = \frac{(2k)!(2k+1)(2k+2)}{2^k k! 2(k+1)} = M(2k+1),$$

where M is an integer by the induction assumption. Hence, the assertion is also true for $k+1$, completing the argument.

13. Clearly $a_1 = 1 < 2^1$, $a_2 = 2 < 2^2$ and $a_3 = 3 < 2^3$. For the induction step, suppose that $a_k < 2^k$ for all integers $1 \leq k < n$. It follows that

$$\begin{aligned}
a_n = a_{n-1} + a_{n-2} + a_{n-3} &< 2^{n-1} + 2^{n-2} + 2^{n-3} \\
&= 2^{n-3}(4+2+1) \\
&= 7 \cdot 2^{n-3} < 8 \cdot 2^{n-3} = 2^n.
\end{aligned}$$

By the Second Principle of Finite Induction, $a_n < 2^n$ for all positive integers n.

1.2 The Binomial Theorem

1. (a) For $n \geq k \geq r \geq 0$,

$$\begin{aligned}
\binom{n}{k}\binom{k}{r} &= \frac{n!}{k!(n-k)!} \cdot \frac{k!}{r!(k-r)!} \\
&= \frac{n!}{r!(n-k)!(k-r)!} \\
&= \frac{n!}{r!(n-r)!} \cdot \frac{(n-r)!}{(k-r)!(n-k)!} \\
&= \binom{n}{r}\binom{n-r}{k-r}.
\end{aligned}$$

(b) Take $r = 1$ in part (a) to obtain

$$\binom{n}{k} k = n \binom{n-1}{k-1} = n \cdot \frac{(n-1)!}{(k-1)!(n-k)!}$$

$$= (n-k+1) \cdot \frac{n!}{(k-1)!(n-k+1)!}$$

$$= (n-k+1) \binom{n}{k-1}.$$

3. (a) Taking $a = b = 1$ in the binomial theorem yields

$$2^n = (1+1)^n = \binom{n}{0} + \binom{n}{1} + \cdots + \binom{n}{n}.$$

(b) Taking $a = 1$ and $b = -1$ in the binomial theorem yields

$$0 = (1-1)^n = \binom{n}{0} - \binom{n}{1} + \cdots + (-1)^n \binom{n}{n}.$$

(c) Since

$$n(1+b)^{n-1} = n\left[\binom{n-1}{0} + \binom{n-1}{1} b + \cdots + \binom{n-1}{n-1} b^{n-1}\right],$$

letting $b = 1$ produces

$$n2^{n-1} = n\binom{n-1}{0} + n\binom{n-1}{1} + \cdots + n\binom{n-1}{n-1}.$$

But

$$n\binom{n-1}{k} = (k+1)\binom{n}{k+1},$$

so that

$$n2^{n-1} = \binom{n}{1} + 2\binom{n}{2} + \cdots + n\binom{n}{n}.$$

(d) Taking $a = 1$ and $b = 2$ in the binomial theorem gives

$$3^n = (1+2)^n = \binom{n}{0} + \binom{n}{1} 2 + \binom{n}{2} 2^2 + \cdots + \binom{n}{n} 2^n.$$

(e) The addition of the expressions in parts (a) and (b) yields

$$2\binom{n}{0} + 2\binom{n}{2} + 2\binom{n}{4} + \cdots = 2^n + 0,$$

while subtraction produces

$$2\binom{n}{1} + 2\binom{n}{3} + 2\binom{n}{5} + \cdots = 2^n - 0.$$

(f) The hint in part (c) implies that

$$\frac{1}{k+1}\binom{n}{k} = \frac{1}{n+1}\binom{n+1}{k+1}.$$

Thus

$$\binom{n}{0} - \frac{1}{2}\binom{n}{1} + \frac{1}{3}\binom{n}{2} + \cdots + \frac{(-1)^n}{n+1}\binom{n}{n}$$

$$= \frac{1}{n+1}\left[\binom{n+1}{1} - \binom{n+1}{2} + \binom{n+1}{3}\right.$$

$$\left. + \cdots + (-1)^n\binom{n+1}{n+1}\right]$$

$$= \frac{1}{n+1}\binom{n+1}{0}$$

$$= \frac{1}{n+1},$$

using part (b).

5. (a) If $n = 2$, then $\binom{2}{2} = 1 = \binom{3}{3}$ so the assertion holds in this case. Suppose that for some integer k we have

$$\binom{2}{2} + \binom{3}{2} + \cdots + \binom{k}{2} = \binom{k+1}{3}.$$

Then

$$\binom{2}{2} + \binom{3}{2} + \cdots + \binom{k}{2} + \binom{k+1}{2}$$

$$= \binom{k+1}{3} + \binom{k+1}{2} = \binom{k+2}{3}.$$

Hence the assertion is true for $k + 1$, completing the induction.

(b) For $n \geq 1$,

$$
\begin{aligned}
1^2 + 2^2 &+ 3^2 + \cdots + n^2 \\
&= 1 + 2\binom{2}{2} + 2 + 2\binom{3}{2} + 3 + \cdots + 2\binom{n}{2} + n \\
&= (1 + 2 + 3 + \cdots + n) + 2\left[\binom{2}{2} + \binom{3}{2} + \cdots + \binom{n}{2}\right] \\
&= \frac{n(n+1)}{2} + 2\binom{n+1}{3} \\
&= \frac{n(n+1)}{2} + \frac{(n+1)n(n-1)}{3} = \frac{n(n+1)(2n+1)}{6}.
\end{aligned}
$$

(c) For $n \geq 2$,

$$
1 \cdot 2 + 2 \cdot 3 + \cdots + n(n+1) = 2\binom{2}{2} + 2\binom{3}{2} + \cdots + 2\binom{n+1}{2}
$$

$$
= 2\binom{n+2}{3} = \frac{n(n+1)(n+2)}{3}.
$$

7. Use induction. When $n = 1$, $1^2 = \binom{3}{3}$ making the assertion correct for $n = 1$. Assume that

$$
1^2 + 3^2 + 5^2 + \cdots + (2k-1)^2 = \binom{2k+1}{3}
$$

for some integer k. Then

$$
\begin{aligned}
1^2 + 3^2 &+ 5^2 + \cdots + (2k-1)^2 + (2k+1)^2 \\
&= \binom{2k+1}{3} + (2k+1)^2 \\
&= \frac{(2k-1)2k(2k+1)}{6} + (2k+1)^2 \\
&= \frac{(2k+1)(2k+2)(2k+3)}{6} = \binom{2k+3}{3}.
\end{aligned}
$$

Thus the truth of the assertion for k implies its truth for $k+1$, completing the induction step.

9. Since $2k > 2k - 1 > k$ for $k = 1, 2, \cdots n$, we have

$$2 \cdot 4 \cdot 6 \cdots (2n) > 1 \cdot 3 \cdot 5 \cdots (2n - 1) > 1 \cdot 2 \cdot 3 \cdots n.$$

Upon multiplying by $2^n n!$, this inequality becomes

$$(2 \cdot 4 \cdot 6 \cdots (2n)) 2^n n! > (2n)! > 2^n (n!)^2,$$

hence division by $(n!)^2$ yields

$$2^{2n} > \binom{2n}{n} > 2^n.$$

Chapter 2

Divisibility Theory in the Integers

2.1 Early Number Theory

1. (a) A number t is triangular if and only if $t = 1 + 2 + 3 + \cdots + n$ for some $n \geq 1$. By Problem 1(a) of Section 1.1, $1 + 2 + 3 + \cdots n = n(n+1)/2$.

   (b) Suppose n is triangular, that is, $n = k(k+1)/2$ for some $k \geq 1$. Then

   $$8n + 1 = 4k(k+1) + 1 = (2k+1)^2,$$

   a perfect square. Conversely, if $8n + 1 = r^2$, then r must be an odd integer; say $r = 2s + 1$. Then $8n + 1 = (2s+1)^2$ leads to $n = s(s+1)/2$, so n is triangular.

   (c) For $n \geq 2$,

   $$t_{n-1} + t_n = \frac{(n-1)n}{2} + \frac{n(n+1)}{2} = n^2.$$

   (d) If $n = t_k = k(k+1)/2$, it follows that

   $$
   \begin{aligned}
   9n + 1 &= \frac{9k(k+1)}{2} + 1 = \frac{(3k+1)(3k+2)}{2} = t_{3k+1}, \\
   25n + 3 &= \frac{25k(k+1)}{2} + 3 = \frac{(5k+2)(5k+3)}{2} = t_{5k+3}, \\
   49n + 6 &= \frac{49k(k+1)}{2} + 6 = \frac{(7k+3)(7k+4)}{2} = t_{7k+3}.
   \end{aligned}
   $$

3. If n is even, say $n = 2k$, then

$$
\begin{aligned}
t_1 + t_2 + \cdots + t_n &= (t_1 + t_2) + (t_3 + t_4) + \cdots + (t_{2k-1} + t_{2k}) \\
&= 2^2 + 4^2 + \cdots + (2k)^2 \\
&= 4(1^2 + 2^2 + \cdots + k^2) \\
&= 4\frac{k(2k+1)(k+1)}{6} = \frac{n(n+1)(n+2)}{6}.
\end{aligned}
$$

If n is odd, say $n = 2k + 1$, then

$$
\begin{aligned}
t_1 + t_2 + \cdots + t_n &= (t_1 + t_2) + (t_3 + t_4) + \cdots + (t_{2k-1} + t_{2k}) + t_{2k+1} \\
&= 4\frac{k(2k+1)(k+1)}{6} + \frac{(2k+1)(2k+2)}{2} \\
&= \frac{(2k)(2k+1)(k+2)}{6} + \frac{3(2k+1)(2k+2)}{6} \\
&= \frac{(2k+1)(2k+2)(2k+3)}{6} = \frac{n(n+1)(n+2)}{6}.
\end{aligned}
$$

7. For $n \geq 2$,

$$
t_n^2 - t_{n-1}^2 = \frac{n^2(n+1)^2}{4} - \frac{(n-1)^2 n^2}{4} = n^3.
$$

9. If $x = \frac{n(n+3)}{2} + 1$, $y = n + 1$ and $z = n(n+3)/2$, then

$$
\begin{aligned}
t_y + t_z &= \frac{(n+1)(n+2)}{2} + \frac{1}{2}\left[\frac{n(n+3)}{2}\right]\left[\frac{n(n+3)}{2} + 1\right] \\
&= \frac{1}{8}[4(n+1)(n+2) + n(n+3)(n(n+3) + 4)] \\
&= \frac{1}{8}[n(n+3) + 2][n(n+3) + 4] \\
&= \frac{1}{2}\left[\frac{n(n+3)}{2} + 1\right]\left[\frac{n(n+3)}{2} + 2\right] = t_x.
\end{aligned}
$$

11. (a) For $n \geq 2$,

$$
t_{n-1} + n^2 = \frac{(n-1)n + 2n^2}{2} = \frac{n(3n-1)}{2} = p_n.
$$

(b) For $n \geq 2$,

$$
3t_{n-1} + n = \frac{3(n-1)n + 2n}{2} = \frac{n(3n-1)}{2} = p_n.
$$

Also,

$$
2t_{n-1} + t_n = \frac{2(n-1)n + n(n+1)}{2} = \frac{n(3n-1)}{2} = p_n.
$$

2.2   The Division Algorithm

1. There exist unique integers q' and r' for which $a = q'b + r'$, $0 \le r' < b$. Thus $a = (q' - 2)b + (r' + 2b) = qb + r$, where $q = q' - 2$ and $r = r' + 2b$ satisfies $2b \le r < 3b$.

3. (a) Any integer n has the form $n = 3q + r$, where $0 \le r \le 2$. Considering cases:

$$(3q)^2 = 3(3q^2) = 3k,$$
$$(3q + 1)^2 = 3(3q^2 + 2q) + 1 = 3k + 1,$$
$$(3q + 2)^2 = 3(3q^2 + 4q + 1) + 1 = 3k + 1.$$

   (b) Write $n = 3q + r$, $0 \le r \le 2$, and consider cases:

$$(3q)^3 = 9(3q^3) = 9k,$$
$$(3q + 1)^3 = 9(3q^3 + 3q^2 + q) + 1 = 9k + 1,$$
$$(3q + 2)^3 = 9(3q^3 + 6q^2 + 3q) + 8 = 9k + 8.$$

   (c) Any integer n is of the form $n = 5q + r$, where $0 \le r \le 4$. Considering cases,

$$(5q)^4 = 5(125q^4) = 5k,$$
$$(5q + 1)^4 = [5q(5q + 2) + 1]^2 = 5k + 1,$$
$$(5q + 2)^4 = [5q(5q + 4) + 4]^2 = 5k + 1,$$
$$(5q + 3)^4 = [5q(5q + 6) + 9]^2 = 5k + 1,$$
$$(5q + 4)^4 = [5q(5q + 8) + 16]^2 = 5k + 1.$$

5. Any positive integer n is of the form $n = 6k + 4$, $0 \le r \le 5$. Considering cases, $n(n + 1)(2n + 1)/6$ is always an integer:

$$(6k)(6k + 1)(12k + 1)/6 = k(6k + 1)(12k + 1),$$
$$(6k + 1)(6k + 2)(12k + 3)/6 = (6k + 1)(3k + 1)(4k + 1),$$
$$(6k + 2)(6k + 3)(12k + 5)/6 = (3k + 1)(2k + 1)(12k + 5),$$
$$(6k + 3)(6k + 4)(12k + 7)/6 = (2k + 1)(3k + 2)(12k + 7),$$
$$(6k + 4)(6k + 5)(12k + 9)/6 = (3k + 2)(6k + 5)(4k + 3),$$
$$(6k + 5)(6k + 6)(12k + 11)/6 = (6k + 5)(k + 1)(12k + 11).$$

9. Any integer is of the form $n = 7q + r$, $0 \le r \le 6$. Considering cases:

$$(7q)^3 = 7(49q^3) = 7k,$$
$$(7q + 1)^3 = [7q(7q + 2) + 1](7q + 1) = 7k + 1,$$

$$(7q+2)^3 = [7q(7q+4)+4](7q+2) = 7k+1,$$
$$(7q+3)^3 = [7(7q^2+6q+1)+2](7q+3) = 7k-1,$$
$$(7q+4)^3 = [7(7q^2+8q+2)+2](7q+4) = 7k+1,$$
$$(7q+5)^3 = [7(7q^2+10q+3)+4](7q+5) = 7k-1,$$
$$(7q+6)^3 = [7(7q^2+12q+5)+1](7q+6) = 7k-1.$$

11. An odd integer n is of the form $4q+1$ or $4q+3$. Considering cases, $n^4 + 4n^2 + 11$ can be expressed as

$$(4q+1)^4+4(4q+1)^2+11 = [8q(2q+1)+1]^2+4[8q(2q+1)+1]+11 = 16k$$

or as

$$(4q+3)^4+4(4q+3)^2+11 = [8(2q^2+3q+1)+1]^2+4[8(2q^2+3q+1)+1]+11 = 16k.$$

2.3   The Greatest Common Divisor

1. If $a|b$, then $b = ac$ for some integer c. Hence

$$b = (-a)(-c), \quad -b = a(-c), \quad -b = (-a)c,$$

which say, respectively, that $(-a)|b$, $a|(-b)$ and $(-a)|(-b)$.

5. Any integer is of the form $a = 3k + r$, $0 \le r \le 2$. Considering cases, we have

(3k)(3k +2)(3k+ 4)=3[k(3k+2)(3k + 4)],
(3k +1)(3k+3)(3k+5)=3[(3k+1)(k + 1)(3k +5)],
(3k +2)(3k+4)(3k+6)=3[(3k+2)(3k + 4)(k +2)].

7. If a and b are both odd integers, then $a^2 = 8k + 1$, $b^2 = 8j + 1$ for some k, j. Thus

$$\begin{aligned} a^4 + b^4 - 2 &= (8k+1)^2 + (8j+1)^2 - 2 \\ &= 16(4k^2 + k + 4j^2 + j). \end{aligned}$$

9. Clearly $N = (a+1)^3 - a^3 = 3a(a+1) + 1$. If $2|n$, then since $2|a(a+1)$ we would have $2|n - 3a(a+1)$ or $2|1$.

13. Put $d = \gcd(a, b)$, so that $d|a$ and $d|b$.

(a) If $c = ax + by$ for some x, y, then $d|ax + by$ or $d|c$. Conversely, suppose $d|c$, say $c = dr$. We know $d = au + bv$ for suitable u, v; hence $c = dr = a(ru) + b(rv) = ax + by$.

(b) Assume that $ax + by = d$. Then $(a/d)x + (b/d)y = 1$, whence $\gcd(x, y) = 1$ by Theorem 2-4.

15. Let $d = \gcd(2a - 3b, 4a - 5b)$, so that $d|2a - 3b$ and $d|4a - 5b$. Then $d|(-2)(2a - 3b) + (4a - 5b)$ or $d|b$. In particular, when $b = -1$, $d|-1$ implies that $d = 1$, hence $\gcd(2a + 3, 4a + 5) = 1$.

17. Since

$$\binom{2n}{n}(2n + 1) = \binom{2n + 1}{n + 1}(n + 1),$$

it follows that $(n + 1)|\binom{2n}{n}(2n + 1)$. But $\gcd(n + 1, 2n + 1) = 1$,

so that $(n + 1)|\binom{2n}{n}$. Therefore, $\frac{1}{n+1}\binom{2n}{n}$ is an integer.

19. (a) Any integer is of the form $a = 6k + r$, $0 \le r \le 5$. Thus, in the various cases, $a(a^2 + 11)$ becomes

$(6k)[(6k)^2 + 11] = 6 \cdot k(36k^2 + 11),$
$(6k + 1)[(6k + 1)^2 + 11] = (6k + 1) \cdot 6(6k^2 + 2k + 2),$
$(6k + 2)[(6k + 2)^2 + 11] = 2(3k + 1) \cdot 3(12k^2 + 8k + 5),$
$(6k + 3)[(6k + 3)^2 + 11] = 3(2k + 1) \cdot 2(18k^2 + 18k + 10),$
$(6k + 4)[(6k + 4)^2 + 11] = 2(3k + 2) \cdot 3(12k^2 + 16k + 9),$
$(6k + 5)[(6k + 5)^2 + 11] = (6k + 5) \cdot 6(6k^2 + 10k + 6).$

(b) If a is odd, then $a^2 = 8k + 1$ for some k. Hence $8|a(a^2 - 1)$, while 3 divides $a(a^2 - 1) = (a - 1)a(a + 1)$, so the product is divisible by 24.

(c) If a and b are both odd, then $a^2 = 8k + 1$, $b^2 = 8j + 1$. Thus $a^2 - b^2 = 8(k - j)$.

(d) If a is not divisible by 2 or 3, then a is of the form $12k + 1$, $12k + 5$, $12k + 7$, or $12k + 11$. Considering cases, $a^2 + 23$ becomes:

$(12k + 1)^2 + 23 = 24(6k^2 + k + 1),$
$(12k + 5)^2 + 23 = 24(6k^2 + 5k + 2),$
$(12k + 7)^2 + 23 = 24(6k^2 + 6k + 3),$
$(12k + 11)^2 + 23 = 24(6k^2 + 11k + 6).$

(e) The product $(a-2)(a-1)a(a+1)$ is divisible by 8, while $(a-2)(a-1)a(a+1)(a+3)$ is divisible by 5; also 3 divides each of $(a-2)(a-1)a$ and $a(a+1)(a+2)$. Thus $360 = 5 \cdot 8 \cdot 9$ divides the product $(a-2)(a-1)a^2(a+1)(a+2)$.

21. (a) If $d|n$, say $n = dr$, then

$$2^n - 1 = 2^{dr} - 1 = (2^d - 1)(2^{d(k-1)} + 2^{d(k-2)} + \cdots + 2^d + 1)$$

so that $2^d - 1 | 2^n - 1$.

(b) Since $5|35$ and $7|35$, $2^5 - 1 | 2^{35} - 1$ and $2^7 - 1 | 2^{35} - 1$.

23. If $d = \gcd(a, b)$ and $d' = \gcd(a, c)$, then $d = ax + by$ and $d' = au + cv$ for suitable x, y, u, v. Hence $dd' = a(axu + byv + cxv) + (bc)(yv)$. Thus if $a|bc$, then $a|dd'$.

2.4   The Euclidean Algorithm

1. To find $\gcd(143, 227)$, we note that

$$
\begin{aligned}
227 &= 1 \cdot 143 + 84 \\
143 &= 1 \cdot 84 + 59 \\
84 &= 1 \cdot 59 + 24 \\
59 &= 2 \cdot 24 + 11 \\
24 &= 2 \cdot 11 + 2 \\
11 &= 5 \cdot 2 + 1 \\
2 &= 2 \cdot 1 + 0
\end{aligned}
$$

Hence, $\gcd(143, 227) = 1$.
To find $\gcd(306, 657)$, we note that

$$
\begin{aligned}
657 &= 2 \cdot 306 + 45 \\
306 &= 6 \cdot 45 + 36 \\
45 &= 1 \cdot 36 + 9 \\
36 &= 4 \cdot 9 + 0
\end{aligned}
$$

Hence, $\gcd(306, 657) = 9$.
To find $\gcd(272, 1479)$ we note that

$$1479 = 5 \cdot 272 + 119$$

$$272 = 2 \cdot 119 + 34$$
$$119 = 3 \cdot 34 + 17$$
$$34 = 2 \cdot 17 + 0$$

Hence, $\gcd(272, 1479) = 17$.

5. (a) Use induction to show that if $\gcd(a, b) = 1$, then $\gcd(a, b^n) = 1$. Assuming $\gcd(a, b^k) = 1$ for some k, the induction is completed by invoking Problem 20(a) of Section 2.3 to conclude that $\gcd(a, bb^k) = 1$ or $\gcd(a, b^{k+1}) = 1$. From the same problem, $\gcd(a, b^n) = 1$ implies that $\gcd(a^m, b^n) = 1$ by induction on m.

   (b) Let $d = \gcd(a, b)$, so that $a = dr$, $b = ds$ with $\gcd(r, s) = 1$. Thus $\gcd(r^n, s^n) = 1$ by part (a). Now $a^n | b^n$ implies $r^n | s^n$. Thus $1 = \gcd(r^n, s^n) \geq r^n$ and so $r = 1$. Therefore $a = d = \gcd(a, b)$ and so $a | b$.

9. Let $d = \gcd(a, b)$ and $m = \text{lcm}(a, b)$. Then $a = dr$ and $m = sa$ for some r, s. Hence $m = sa = s(dr) = (sr)d$, so $d | m$.

11. If $d = \gcd(a, b, c)$, $e = \gcd(\gcd(a, b), c)$, then $d | a$, $d | b$, $d | c$; hence $d | \gcd(a, b)$ and $d | c$, so $d \leq e$. On the other hand, $e | \gcd(a, b)$ and $e | c$, which means $e | a$, $e | b$, $e | c$ and so $e \leq d$.

2.5 The Diophantine Equation $ax + by = c$

3. (a) Consider the equation $18x + 5y = 48$. Since $\gcd(18, 5) = 1$ and $1 | 48$, the equation has a solution. Now $1 = 2 \cdot 18 - 7 \cdot 5$ implies that $48 = 96 \cdot 18 - 336 \cdot 5$, whence $x_0 = 96$, $y_0 = -336$ is one solution. The other solutions are $x = 96 + 5t$, $y = -336 - 18t$ for $t = 0, \pm 1, \pm 2, \ldots$. We have $x > 0$, $y > 0$ when $t = -19$, so there is one positive solution $x = 96 - 5 \cdot 19 = 1$, $y = -336 + 18 \cdot 19 = 6$.

   (b) Since $\gcd(54, 21) = 3$ and $3 | 906$, the equation $54x + 21y = 906$ has a solution. Now $3 = 2 \cdot 54 - 5 \cdot 21$ yields $906 = 302 \cdot 3 = 604 \cdot 54 - 1510 \cdot 21$, so that $x_0 = 604$, $y_0 = -1510$. The other solutions are given by

$$x = 604 + 7t,$$
$$y = -1510 - 18t$$

for $t = 0, \pm 1, \pm 2, \ldots$. Because $x > 0$ when $t \geq -86$ and $y > 0$ when $t \leq -84$, positive solutions arise when $t = -86, -85$ or -84 :

$$x = 604 - 7 \cdot 86 = 2, \quad y = -1510 + 18 \cdot 86 = 38;$$
$$x = 604 - 7 \cdot 85 = 9, \quad y = -1510 + 18 \cdot 85 = 20;$$
$$x = 604 - 7 \cdot 84 = 16, \quad y = -1510 + 18 \cdot 84 = 2.$$

(c) Since $\gcd(123, 360) = 3$ and $3|99$, the equation $123x + 360y = 99$ has a solution. The relation $3 = 41 \cdot 123 - 14 \cdot 360$ implies that

$$99 = 33 \cdot 3 = 1353 \cdot 123 - 462 \cdot 360,$$

so that $x_0 = 1353$, $y_0 = -462$ is a solution. The other solutions are

$$x = 1353 + 120t,$$
$$y = -462 - 41t,$$

for $t = 0, \pm 1, \pm 2, \ldots$. Now $x > 0$ when $t \geq -11$ and $y > 0$ when $t \leq -12$, hence there are no positive solutions.

(d) Consider the equation $158x - 57y = 7$. Here, $\gcd(158, -57) = 1$ and $1|7$, so there is a solution. Since $1 = 158(-22) - 57(-61)$, multiplication by 7 gives $7 = 158(-154) - 57(-427)$, so that $x_0 = -154, y_0 = -427$ is a solution. The other solutions of $158x - 57y = 7$ are given by

$$x = -154 - 57t,$$
$$y = -427 - 158t$$

for $t = 0, \pm 1, \pm 2, \ldots$. Positive solutions occur when $t \leq -3$. Replacing t by $s - 3$, the positive solutions are of the form

$$x = 17 - 57s,$$
$$y = 47 - 158s$$

for $s \leq 0$.

5. (a) The equation $10x + 25y = 455$ has the general solution

$$x = -182 + 5t, \quad y = 91 - 2t,$$

with positive solutions occurring when $37 \leq t \leq 45$.

   (b) The equation $180x + 75y = 9000$ has the general solution

$$x = -1200 + 5t, \quad y = 3000 - 12t$$

Nonnegative solutions occur when $240 \leq t \leq 250$, with $x > y$ when $t = 248, 249, 250$.

   (c) The equation $6x + 9y = 126$ implies that $2x + 3y = 42$. This has general solution

$$x = 84 + 3t, \quad y = -42 - 2t.$$

Substitution into $9x + 6y = 114$ gives $t = -26$, hence $x = 6, y = 10$.

7. The equation $100y + x - 68 = 2(100x + y)$ is equivalent to

$$199x - 98y = -68,$$

with general solution

$$x = -2244 - 98t, \quad y = -4556 - 199t.$$

The smallest positive solution (namely, $x = 10$ and $y = 21$) occurs when $t = -23$. Thus the smallest value for which the check could have been written is \$10.21.

Chapter 3

Primes and Their Distribution

3.1 The Fundamental Theorem of Arithmetic

3. (a) If $p = 3n + 1$, then $3n = p - 1$ is even: hence $n = 2m$ for some m, and $p = 6m + 1$.

   (b) If all the prime divisors of $3n + 2$ were of the form $3k + 1$, then their product (namely, $3n + 2$) would also be of this form.

   (c) Suppose the prime $p = n^3 - 1 = (n - 1)(n^2 + n + 1)$. Since this must be the trivial factorization, the smaller factor $n - 1$ will be equal to 1: thus $n = 2$ and $p = 7$.

   (d) Suppose the prime p satisfies $3p + 1 = a^2$ or $3p = a^2 - 1 = (a - 1)(a + 1)$. If $3|a + 1$, say $a + 1 = 3k$, then $p = (3k - 2)k$; one of the factors must be equal to 1 which implies $p = 1$, an impossibility. Thus $3 \nmid a + 1$. Since $3|a - 1$, say $a - 1 = 3k$, we have $p = k(3k + 2)$. Now if $3k + 2 = 1$, then $3k = -1$, which is absurd. So $k = 1$ and $p = 5$.

   (e) Let the prime $p = n^2 - 4 = (n - 2)(n + 2)$. Since this is the trivial factorization, the smaller factor must equal 1; so $n - 2 = 1$ and $p = 5$.

5. (a) If p is a prime and $p|a^n$, then $p|a$; hence $p^n|a^n$.

11. (b) If $k > 2$ is even, say $k = 2n$, with $n \geq 2$, then

$$2^k - 1 = 2^{2n} - 1 = (2^2)^n - 1 = (2^2 - 1)(2^{2(n-1)} + \cdots + 2^2 + 1)$$

is not prime.

13. If $n > 1$ is not of the form $6k + 3$, then it takes one of the forms $6k, 6k + 1, 6k + 2, 6k + 4, 6k + 5$. In the cases $k, 6k + 1$, and $6k + 5$, $3|n^2 + 2n$, while in the remaining cases $2|n^2 + 2^n$.

15. If $a = b^2$, where b has the prime factorization $b = p_1^{k_1} \cdots p_r^{k_r}$ then $a = p_1^{2k_1} \cdots p_r^{2k_r}$. Conversely, if $a = p_1^{2k_1} \cdots p_r^{2k_r}$, then $a = (p_1^{k_1} \cdots p_r^{k_r})^2 = b^2$.

17. Suppose n has the prime factorization $n = 2^k p_1^{k_1} \cdots p_r^{k_r}$, where the p_i are odd primes and $k \geq 0$, $k_j > 0$. Then $n = 2^k m$, with m odd.

19. If n is square-full, the odd exponents in its prime factorization are of the form $2u + 3$. Hence

$$
\begin{aligned}
n &= p_1^{2k_1} \cdots p_r^{2k_r} q_1^{2u_1+3} \cdots q_s^{2u_s+3} \\
&= (p_1^{k_1} \cdots p_r^{k_r} q_1^{u_1} \cdots q_s^{u_s})^2 (q_1 \cdots q_s)^3 = a^2 b^3.
\end{aligned}
$$

3.2   The Sieve of Eratosthenes

3. In the contrary case, n has at least three prime divisors; say, p_1, p_2, p_3 where $p_1 \leq p_2 \leq p_3$. Then $p_1^3 \leq p_1 p_2 p_3 \leq n$, so $p_1 \leq \sqrt[3]{n}$.

5. Suppose $n = ab$ is composite, where $4 \leq n < 1000$. If p, q are primes dividing a and b, respectively, and if p, q are both greater than 31, then $n \geq pq \geq 37^2 > 1000$.

7. Suppose there is a largest prime p and consider $N = p! + 1$. Now N has a prime divisor q and, since $q \leq p$, we have $q|p!$. Thus $q|N - p!$ or $q|1$, which is impossible.

9. (a) For $n > 2$, $n! - 1$ has a prime divisor p. Now $p \neq n!$, for otherwise $p|n! - (n! - 1)$ or $p|1$. Also $p > n$, for otherwise $p|n!$ and so $p|n! - (n! - 1)$. Thus $n < p < n!$.

   (b) Let p be a prime divisor of $p! + 1$. If $p = 2$, then since $2|n!$ for $n > 2$, we would have $2|(n! + 1) - n!$ or $2|1$; thus $p > 2$. Note that $p > n$; for otherwise $p|(n! + 1) - n!$ or $p|1$.

13. (a) Suppose $n|m$, say $m = kn$. Then

$$
9R_m = 10^{kn} - 1 = (10^n - 1)(10^{(k-1)n} + \cdots + 10^n + 1),
$$

   so that $R_m = R_n(10^{(k-1)n} + \cdots + 10^n + 1)$ and $R_n|R_m$.

(b) Note that $9R_{m+n} = 10^{m+n} - 1 = (10^m - 1)10^n + (10^n - 1)$ and so, upon dividing by 9, $R_{m+n} = R_m 10^n + R_n$.

(c) Suppose $\gcd(n, m) = 1$, so that $1 = nr + ms$ for some r, s. Let $d = \gcd(R_n, R_m)$. Then $d|R_n, d|R_m$. By part (a) , $R_n|R_{nr}$ and $R_n|R_{ms}$, whence $d|R_{nr}$ and $d|R_{ms}$. Thus by part (b), $d|R_{nr+ms}$ or $d|R_1$; that is, $d|1$ and so $d = 1$.

3.3   The Goldbach Conjecture

3. Suppose p and q are primes satisfying $p - q = 3$ or $p = q + 3$. Then $q \neq 3$, for otherwise p is composite. If $q > 3$ then q is of the form $6k + 1$ or $6k + 5$; but neither form permits p to be prime. This leaves $q = 2$ and thus $p = 5$.

An alternate and shorter solution: if $p - q = 3$ then one of p and q must be odd and the other even. The only even prime is 2; so q is 2 and p is 5.

9. (a) When $n = 3$, the triplet $n, n + 2, n + 4$ yields the primes 3, 5, 7. Any $n > 3$ takes one of the forms $6q + r$, $0 \leq r \leq 5$; but in each case one of n, $n + 2$, $n + 4$ is composite.

13. Assume the only primes of the form $6k + 5$ are $q_1, q_2, \ldots, q_r$. Consider the integer $N = 6q_1 \cdots q_r - 1 = 6(q_1 \cdots q_r - 1) + 5$ with prime factorization $N = p_1 p_2 \cdots p_s$ where $p_i \neq 2, 3$. If all the p_i were of the form $6k + 1$, then N would be of this form; thus some p_i is of the form $6k + 5$, hence is equal to a q_j. It follows that $p_i|N - 6q_1 \cdots q_r$ or $p_i|1$, which is impossible.

21. (a) If $n = (a + b)(a + 2b) \cdots (a + kb)$, then $a + b$ divides $a + (n + 1)b$, $a + 2b$ divides $(a + n + 2)b$, and so on, with $a + kb$ dividing $a + (n + k)b$.

23. (a) Suppose that

$$\frac{p + (p + 2)}{2} = t_n = \frac{n(n + 1)}{2}.$$

Then $2p + 2 = n(n + 1)$ or $2p = (n + 2)(n - 1)$. If $2|n + 2$, say $n + 2 = 2k$, we get $p = k(2k - 3)$. This implies that $k = 2$ and $p = 2$; but then $p + 2$ will not be prime. It follows that $2|n - 1$, say $n - 1 = 2j$. Then $p = (2j + 3)j$, yielding $j = 1$ and $p = 5$, $p + 2 = 7$.

(b) Suppose that

$$\frac{p + (p + 2)}{2} = a^2,$$

so that $p = (a + 1)(a - 1)$. But because p is prime, this is the trivial factorization: $a + 1 = p$ and $a - 1 = 1$, so that $a = 2$ and $p = 3$.

25. For $n = 4$, the assertion is true since $p_4 = 7 < 2 + 3 + 5$. Assume that for some k, $p_k < p_1 + p_2 + \cdots + p_{k-1}$. By Bertrand's Conjecture there is a prime between p_k and $2p_k$. Hence

$$p_{k+1} \leq 2p_k < (p_1 + p_2 + \ldots + p_{k_1}),$$

and the assertion also holds for $k + 1$.

27. Given $n \geq 2$, suppose n is even, say $n = 2k$. Then there is a prime p satisfying $k < p < 2k = n$: also $n = 2k < 2p$, giving $p < n < 2p$. If n is odd, say $n = 2k + 1$, then $k < p < 2k < 2k + 1 = n$; also $n = 2k + 1 \leq 2p$, giving $p < n \leq 2p$.

Chapter 4

The Theory of Congruences

4.2 Basic Properties of Congruence

1. (a) Suppose $a \equiv b \pmod{n}$, so that $n \mid a - b$. If $m \mid n$, then $m \mid a - b$, hence $a \equiv b \pmod{m}$.

   (b) If $a \equiv b \pmod{n}$, then $n \mid a - b$; thus, for any $c > 0$, $cn \mid c(a - b)$ and so $ca \equiv cb \pmod{cn}$.

   (c) Suppose $a \equiv b \pmod{n}$, so that $a - b = kn$ for some k. If d divides a, b and n, then $a/d - b/d = k(n/d)$; thus $a/d \equiv b/d \pmod{n/d}$.

3. Let $a \equiv b \pmod{n}$, so that $a = b + kn$ for some k. By the lemma in Section 2.3, $\gcd(a, n) = \gcd(b + kn, n) = \gcd(b, n)$.

5. (a) Notice that

$$53^{103} + 103^{53} \equiv (-1)^{103} + 1^{53} \equiv 0 \pmod{3} \text{ and}$$
$$53^{103} + 103^{53} \equiv 1^{103} + (-1)^{53} \equiv 0 \pmod{13}.$$

   Since 3 and 13 both divide $53^{103} + 103^{53}$, with $\gcd(3, 13) = 1$, $39 = 3 \cdot 13$ also divides this sum.

   (b) Modulo 7,

$$111^{333} + 333^{111} \equiv (-1)^{333} + 4^{111}$$
$$\equiv -1 + (4^3)^{37} \equiv -1 + 1^{37} \equiv 0.$$

7. When $n = 1$, $(-13)^2 = 169 \equiv -12 \equiv -13 + 1 \pmod{181}$ so the assertion is true. For some k, assume that

$$(-13)^{k+1} \equiv (-13)^k + (-13)^{k-1} \pmod{181}.$$

23

Then

$$(-13)^{k+2} = (-13)^{k+1} \cdot 13 \equiv [(-13)^k + (-13)^{k+1}]13$$
$$\equiv (-13)^{k+1} + (-13)^k \pmod{181},$$

so the asserted congruence also holds for $k + 1$.

9. For a prime p with $n < p < 2n$,

$$n!\binom{2n}{n} = n(n+1)(n+2)\cdots(2n) \equiv 0 \pmod{p}$$

and so p divides the left side. Since $\gcd(n!, p) = 1$, p must divide $\binom{2n}{n}$.

11. Modulo 11, the integers $0, 1, 2, 2^2, \ldots, 2^9$ are congruent, respectively, to the complete set of residues 0, 1, 2, 4, 8, 5, 10, 9, 7, 3, 6. But $0, 1^2, 2^2, 3^2, \ldots, 10^2$ are congruent, respectively, to 0, 1, 4, 9, 5, 3, 3, 5, 10, 4, 1.

13. Let $a \equiv b \pmod{n_1}$ and $a \equiv b \pmod{n_2}$, so that $a - b = kn_1 = jn_2$ for some k, j. If $d = \gcd(n_1, n_2)$, then $k(n_1/d) = j(n_2/d)$, with $\gcd(n_1/d, n_2/d) = 1$. By Euclid's Lemma, $(n_2/d)|k$, say $k = (n_2/d)r$. Thus $a - b = (n_2/d)rn_1 = (n_1 n_2/d)r$. But $n_1 n_2/d = \text{lcm}(n_1, n_2) = n$ by Theorem 2.8, which yields $a \equiv b \pmod{n}$.

15. Use induction. When $n = 1$, $a^2 \equiv 1 \pmod{23}$ by Problem 8(a). Suppose that for some k we have $a^{2^k} \equiv 1 \pmod{2^{k+1}}$; then $a^{2^k} = 1 + r2^{k+2}$ for some r. On squaring, it follows that

$$a^{2^{k+1}} = (a^{2^k})^2 = (1 + r^{2k+2})^2$$
$$= 1 + r \cdot 2^{k+3} + r^2 2^{2k+4} \equiv 1 \pmod{2k+3},$$

and so the asserted congruence holds for $k + 1$.

17. If $ab \equiv cd \pmod{n}$ and $b \equiv d \pmod{n}$, then $ab \equiv cd \equiv cb \pmod{n}$. But $\gcd(b, n) = 1$, so the last congruence implies that $a \equiv c \pmod{n}$.

4.3 Binary and Decimal Representations of Integers

1. Note that $47 = 32 + 8 + 4 + 2 + 1$. Successive squarings give, modulo 1537:

$$141^2 \equiv 1437, 141^4 \equiv 778, 141^8 \equiv 1243, 141^{16} \equiv 364, 141^{32} \equiv 314.$$

Thus,

$$141^{47} \equiv 314 \cdot 1243 \cdot 778 \cdot 1437 \cdot 141 \equiv 658 \pmod{1537}.$$

For $53 = 32 + 16 + 4 + 1$ we have, modulo 503,

$$19^2 \equiv 361, 19^4 \equiv 44, 19^8 \equiv 427, 19^{16} \equiv 243, 19^{32} \equiv 198.$$

It follows that

$$1653 \equiv 198 \cdot 243 \cdot 44 \cdot 19 \equiv 406 \pmod{503}.$$

3. Since $9t^{10} \equiv 1 \pmod{100}$, $9 \cdot 9^9 \equiv -99 \pmod{100}$ or $9^9 \equiv -11 \equiv 89 \pmod{100}$. Thus

$$9^{9^9} \equiv 9^{9+10k} \equiv (9^9)(9^{10})^k \equiv 89 \cdot 1 \equiv 89 \pmod{100}.$$

5. (a) Let $N = a_m b^m + \cdots + a_1 b + a_0$, where $0 \le a_k \le b - 1$. Since $b \equiv 1 \pmod{b-1}$, it follows that

$$N \equiv a_m + \cdots + a_1 + a_0 \pmod{b-1}.$$

Thus $N \equiv 0 \pmod{b-1}$ if and only if $a_m + \cdots + a_1 + a_0 \equiv 0 \pmod{b-1}$.

   (b) Using part (a), if $b = 9$, then $8|n$ if and only if $8|a_m + \cdots a_1 + a_0$ Also $3|N$ if and only if $3|a_0$.

   (c) If $N = (447836)_9$, then $8|N$ and $3|N$.

7. Suppose that $N = a_m 10^m + \cdots + a_1 10 + a_0$, where $0 \le a_k < 10$.

   (a) $N \equiv a_0 \pmod 2$, so that $N \equiv 0 \pmod 2$ if and only if $a_0 \equiv 0 \pmod 2$.

   (b) $N \equiv a_m + \cdots + a_1 + a_0 \pmod 3$, so that $N \equiv 0 \pmod 3$ if and only if $a_m + \cdots + a_1 + a_0 \equiv 0 \pmod 3$.

    (c) $N \equiv a_1 10 + a_0 \pmod 4$, so that $N \equiv 0 \pmod 4$ if and only if $a_1 10 + a_0 \equiv 0 \pmod 4$.

    (d) $N \equiv a_0 \pmod 5$, so that $N \equiv 0 \pmod 5$, if and only if $a_0 \equiv 0 \pmod 5$.

9. Modulo 9,

$$4444^{4444} \equiv (-2)^{4444} \equiv 2(2^3)^{1481} \equiv 2(-1)^{1481} \equiv -2 \equiv 7.$$

11. If $5 \cdot 9 \cdot 11 | 273x49y5$, then $x + y \equiv 6 \pmod 9$ and $x - y \equiv -1 \pmod{11}$; hence, $x = 7, y = 8$.

13. Note that $t_{n+2k} - t_n = k[2n + (2k + 1)] \equiv 0 \pmod k$; in particular $t_{n+20} \equiv t_n \pmod{10}$, so that each has the same last digit.

15. For $n \geq 5$, $1! + 2! + 3! + \cdots + n! \equiv 1! + 2! + 3! + 4! \equiv 3 \pmod{10}$; hence, the expression cannot be a square by Problem 2(a). Squares occur when $n = 1$ or 3, but not when $n = 2$ or 4.

17. Let $N = a_m 10^m + \cdots + a_1 10 + a_0$.

    (a) Modulo 1001, if k is even, then $10^{3k} \equiv 1$, $10^{3k+1} \equiv 10$, $10^{3k+2} \equiv 100$; if k is odd, then $10^{3k} \equiv -1$, $10^{3k+1} \equiv -10$, $10^{3k+2} \equiv -100$. Thus

$$
\begin{aligned}
N \equiv\ & (a_0 + 10a_1 + 100a_2) - (a_3 + 10a_4 + 100a_5) \\
& + (a_6 + 10a_7 + 100a_8) - \cdots.
\end{aligned}
$$

So $N \equiv 0 \pmod{1001}$ if and only if the above expression is congruent to 0 modulo $1001 = 7 \cdot 11 \cdot 13$.

    (b) Since $10 \equiv 4 \pmod 6$ implies that $10^k \equiv 4^k \equiv 4 \pmod 6$ for $k \geq 1$,
$$N \equiv 4a_m + \cdots + 4a_2 + 4a_1 + a_0 \pmod 6.$$

Thus $N \equiv 0 \pmod 6$ if and only if

$$4a_m + \cdots + 4a_2 + 4a_1 + a_0 \equiv 0 \pmod 6.$$

19. (a) If $N = a_m 10^m + am_{m-1} 10^{m-1} + \cdots + a_1 10 + a_0$ and $M = a_0 10^m + a_1 10^{m-1} + \cdots + a_{m-1} 10 + a_m$, then

$$
\begin{aligned}
N - M \equiv\ & (a_m + a_{m-1} + \cdots + a_1 + a_0) - (a_0 + a_1 + \cdots + a_{m-1} + a_m) \\
\equiv\ & 0 \pmod 9.
\end{aligned}
$$

(b) If $N = a_0 10^{2n+1} + a_1 l0^{2n} + \cdots + a_1 10 + a_0$, then

$$N \equiv a_0 - a_1 + \cdots + a_1 - a_0 \equiv 0 \pmod{11}$$

hence $11|N$

21. By Problem 17(a), 7, 11 and 13 all divide R_6; also $3|R_6$ since $3|6$. The remaining factor of R_6 is 37.

23. Since $72 = 9 \cdot 8$ divides $x679y$, $x + y \equiv 5 \pmod 9$. Also $8|79y$, so that $y = 2$; consequently, $x = 3$.

25. Any prime $p > 3$ is of the form $6k + 1$ or $6k + 5$. When $p = 6k + 1$, the congruence $10^6 \equiv 1 \pmod{13}$ implies that

$$
\begin{aligned}
10^{2p} - 10^p + 1 &\equiv (10^6)^{2k} 10^2 - (10^6)^k 10 + 1 \\
&\equiv 100 - 10 + 1 \equiv 0 \pmod{13}.
\end{aligned}
$$

A similar argument holds in the case $p = 6k + 5$.

27. (a) The number is correct, since the check digit

$$
\begin{aligned}
a_{10} &\equiv 1 \cdot 0 + 2 \cdot 0 + 3 \cdot 7 + 4 \cdot 2 + 5 \cdot 3 + 6 \cdot 2 + 7 \cdot 5 + 8 \cdot 6 + 9 \cdot 9 \\
&\equiv 220 \equiv 0 \pmod{11}.
\end{aligned}
$$

(b) The number is incorrect, since the check digit

$$
\begin{aligned}
a_{10} &\equiv 1 \cdot 9 + 2 \cdot 5 + 3 \cdot 7 + 4 \cdot 6 + 5 \cdot 4 + 6 \cdot 3 + 7 \cdot 4 + 8 \cdot 9 + 9 \cdot 7 \\
&\equiv 257 \equiv 4 \pmod{11}, \text{ not } 5.
\end{aligned}
$$

(c) The number is correct, since the check digit

$$
\begin{aligned}
a_{10} &\equiv 1 \cdot 1 + 2 \cdot 5 + 3 \cdot 6 + 4 \cdot 9 + 5 \cdot 4 + 6 \cdot 7 + 7 \cdot 3 + 8 \cdot 0 + 9 \cdot 3 \\
&\equiv 175 \equiv 10 \pmod{11}.
\end{aligned}
$$

4.4   Linear Congruences and the Chinese Remainder Theorem

1. (a) If $25x \equiv 15 \pmod{29}$, then $5x \equiv 3 \equiv -55 \pmod{29}$; hence $x \equiv -11 \equiv 18 \pmod{29}$.

(b) If $5x \equiv 2 \pmod{26}$, then $5x \equiv -50 \pmod{26}$; hence $x \equiv -10 \equiv 16 \pmod{26}$.

(c) If $6x \equiv 15 \pmod{21}$, then $2x \equiv 5 \equiv 12 \pmod 7$; hence $x \equiv 6 \pmod 7$, so $x \equiv 6, 13, 20 \pmod{21}$.

(d) If $36x \equiv 8 \pmod{102}$, then there is no solution, since $\gcd(36, 102) \equiv 6$ does not divide 8.

(e) If $34x \equiv 60 \pmod{98}$, then $17x \equiv 30 \pmod{49}$ or $17 \equiv 30 - 98 \equiv -68 \pmod{49}$: thus $x \equiv -4 \equiv 45 \pmod{49}$, and so $x \equiv 45, 94 \pmod{98}$.

(f) If $140x \equiv 133 \pmod{301}$, then $20x \equiv 19 \equiv 62 \pmod{43}$. Thus $10x \equiv 31 \equiv 74 \pmod{43}$, so $5x \equiv 37 \equiv 80 \pmod{43}$; it follows that $x \equiv 16 \pmod{43}$ and so

$$x \equiv 16, 59, 102, 145, 188, 231, 274 \pmod{301}.$$

5. The system of congruences

$$17x \equiv 3 \pmod 2 \qquad 17x \equiv 3 \pmod 3$$
$$17x \equiv 3 \pmod 5 \qquad 17x \equiv 3 \pmod 7$$

is equivalent to the system

$$x \equiv 1 \pmod 2 \qquad 2x \equiv 0 \pmod 3$$
$$2x \equiv 3 \pmod 5 \qquad 3x \equiv 3 \pmod 7$$

with individual solutions

$x \equiv 1 \pmod 2$, $x \equiv 0 \pmod 3$, $x \equiv 4 \pmod 5$, and $x \equiv 1 \pmod 7$.

The first of the congruences implies $x = 1 + 2t$. For the second to hold $t = 1 + 3s$, so that $x = 3 + 6s$. Substitution into the third congruence gives $s = 1 + 5r$, whence $x = 9 + 30r$. The last congruence yields $r = 3 + 7k$ and $x = 99 + 210k$.

7. (a) In the system of congruences

$$a \equiv 0 \pmod 4, \ a \equiv -1 \pmod 9, \ a \equiv -2 \pmod{25},$$

the first has solution $a = 4t$. For the second to hold, $t = 2 + 9r$ so that $a = 8 + 36r$. The last congruence implies that $r = 15 + 25k$, yielding $a = 548 + 900k$.

(b) Given the system

$$a \equiv 0 \pmod{25}, \ a \equiv -1 \pmod{27}, \ a \equiv -2 \pmod{16},$$

the first implies that $a = 25t$. Substitution into the second congruence yields $t = 14 + 27r$ and so $a = 350 + 25 - 27r$. The remaining congruence gives $r = 16k$; hence, $a = 350 + 25 \cdot 27 \cdot 16k$.

9. The system of congruences

$$x \equiv 1 \pmod{2}, \qquad x \equiv 1 \pmod{3}, \qquad x \equiv 1 \pmod{4},$$
$$x \equiv 1 \pmod{5}, \qquad x \equiv 1 \pmod{6}, \qquad x \equiv 0 \pmod{7}$$

is equivalent to the system

$$x \equiv 1 \pmod{3}, \ x \equiv 1 \pmod{4}, \ x \equiv 1 \pmod{5}, \ x \equiv 0 \pmod{7}$$

The first of these has solution $x = 1 + 3t$ and the second implies that $t = 4r$, whence $x = 1 + 12r$. Substituting into the third congruence yields $r = 5s$, so that $x = 1 + 60s$. The last congruence gives $s = 5 + 7k$. Thus $x = 301 + 420k$, with smallest positive value 301.

11. If $x \equiv a \pmod{n}$, then $x = a + kn$ for some k. Substituting into the congruence $x \equiv b \pmod{m}$, we get $kn \equiv b - a \pmod{m}$. This linear congruence has a solution of k if and only if $\gcd(n, m) | b - a$. If x_0 and y_0 are two solutions of $x \equiv a \pmod{n}$, $x \equiv b \pmod{m}$, then $x_0 \equiv a \equiv y_0 \pmod{n}$, $x_0 \equiv b \equiv y_0 \pmod{m}$; hence $n | x_0 - y_0$ and $m | x_0 - y_0$, which makes $x_0 - y_0$ a common multiple of n and m. By Problem 10(c) of Section 2.4, $\mathrm{lcm}(n, m) | x_0 - y_0$ and so x_0 and y_0 are congruent modulo $\mathrm{lcm}(n, m)$.

13. Since $x \equiv a \pmod{n}$, $x - a = kn$ for some k. If k is even, it follows that $x \equiv a \pmod{2n}$, while if k is odd, then $x \equiv a + n \pmod{2n}$.

15. (a) The system of congruences

$$x \equiv 1 \pmod{2}, \ x \equiv 2 \pmod{3}, \ x \equiv 5 \pmod{6}, \ x \equiv 5 \pmod{12}$$

is equivalent to the system

$$x \equiv 1 \pmod{2}, \ x \equiv 2 \pmod{3}, \ x \equiv 5 \pmod{12}.$$

The first of these implies that $x = 1 + 2t$, while the second gives $t = 2 + 3r$; hence, $x = 5 + 6r$. The last congruence yields $r = 2k$, so that $x = 5 + 12k$. One solution is $x = 17$.

(b) The system

$$x \equiv 2 \pmod 3, \ x \equiv 3 \pmod 4, \ x \equiv 4 \pmod 5, \ x \equiv 5 \pmod 6$$

can be reduced to

$$x \equiv 3 \pmod 4, \ x \equiv 4 \pmod 5, \ x \equiv 5 \pmod 6$$

The first congruence has solution $x = 3 + 4t$ and the second implies that $t = 4 + 5r$; thus $x = 19 + 20r$. Substitution into the last congruence yields $r = 2 + 3k$, so that $x = 59 + 60k$.

(c) In the system

$$x \equiv 3 \pmod{10}, \ x \equiv 11 \pmod{13}, \ x \equiv 15 \pmod{17},$$

the first congruence gives $x = 3 + 10t$. The second implies that $t = 6 + 13r$ and so $x = 63 + 130r$. From the last congruence, it follows that $r = 8 + 17k$, giving $x = 1103 + 2210k$.

17. Multiplying the equations in the system

$$3x + 4y \equiv 5 \pmod{13}, \ 2x + 5y = 7 \pmod{13}$$

by 2 and 3, respectively, and then subtracting the results leads to $6y \equiv 2 \pmod{13}$; thus $y \equiv 9 \pmod{13}$. Substitution into either equation of the original system produces $x \equiv 7 \pmod{13}$.

19. The congruence $3x + 4y \equiv 5 \pmod 8$ may be written as $3x \equiv 5 - 4y$ $\pmod 8$. Since $\gcd(3, 8) = 1$, each of the eight incongruent values of y will lead to a unique value of x. Hence, there are eight incongruent solutions modulo 8. These are:

$$x \equiv 7, y \equiv 0; \qquad x \equiv 3, y \equiv 1; \qquad x \equiv 7, y \equiv 2;$$
$$x \equiv 3, y \equiv 3; \qquad x \equiv 7, y \equiv -4; \qquad x \equiv 3, y \equiv 5;$$
$$x \equiv 7, y \equiv 6; \qquad x \equiv 3, y \equiv 7.$$

Chapter 5

Fermat's Theorem

5.2 Fermat's Little Theorem and Pseudoprimes

1. Modulo 17,

$$11^{104} \equiv (11^2)^{52} \;\equiv\; 2^{52} \equiv (2^5)^{10} 2^2 \equiv (-2)^{10} \cdot 4$$
$$\equiv\; 4 \cdot 4 \equiv 16 \equiv -1.$$

3. Modulo 13, $1 + 11^{12n+6} \equiv 1 + (11^{12})^n \cdot 11^6 \equiv 1 + 1(-2)^6 \equiv 0$.

5. Let $\gcd(a, 30) = 1$. Since $\gcd(a, 5) = \gcd(a, 3) = 1$, $a^4 \equiv 1 \pmod{5}$ and $a^2 \equiv 1 \pmod{3}$, whence $a^4 \equiv 1 \pmod{3}$. Now a is an odd integer, so that $a^2 = 8k + 1$ for some k; in particular, $a^2 \equiv 1 \pmod{4}$, so that $a^4 \equiv 1 \pmod{4}$. Because 3, 4 and 5 all divide $a^4 - 1$, it follows that $3 \cdot 4 \cdot 5 | (a^4 - 1) + 60$.

7. Since $\gcd(a, 7) = 1$, $a^6 \equiv 1 \pmod{7}$ or $7 | (a^3 - 1)(a^3 + 1)$. It follows that $7 | a^3 - 1$ or $7 | a^3 + 1$.

9. (a) $x \equiv a^{p-2} b \pmod{p}$ satisfies $ax \equiv b \pmod{p}$, since

    $$a(a^{p-2}b) \equiv a^{p-1} b \equiv 1 \cdot b \equiv b \pmod{p}.$$

   (b) By part (a), $2x \equiv 1 \pmod{31}$ is satisfied by $x \equiv 2^{29} \equiv (2^5)^5 2^4 \equiv 16 \pmod{31}$;
   $6x \equiv 5 \pmod{11}$ is satisfied by $x \equiv 6^9 \cdot 5 \equiv 10 \pmod{11}$;
   and $3x \equiv 17 \pmod{29}$ is satisfied by $x \equiv 3^{27} \cdot 17 \equiv (3^3)^9 \cdot 17 \equiv 25 \pmod{29}$.

11. Because $\gcd(a,p) = 1$ for $1 \le a \le p-1$, $a^{p-1} \equiv 1 \pmod{p}$. Then:

   (a) $1^{p-1} + 2p^{p-1} + \cdots + (p-1)^{p-1} \equiv 1 + 1 + \cdots + 1 \equiv p - 1 \equiv -1 \pmod{p}$;

   (b) $1^p + 2^p + \cdots + (p-1)^p \equiv 1 + 2 + \cdots + (p-1) \equiv \frac{(p-1)p}{2} \equiv kp \equiv 0 \pmod{p}$.

13. Let $\gcd(a,pq) = 1$. Since $\gcd(a,q) = 1$, $a^{q-1} \equiv 1 \pmod{q}$. Also $\gcd(a,p) = 1$ implies that $a^{p-1} \equiv 1 \pmod{p}$. But $p-1 \mid q-1$; say, $q-1 = k(p-1)$. Thus $a^{q-1} \equiv (a^{p-1})^k \equiv 1 \pmod{p}$. These congruences give $a^{q-1} \equiv 1 \pmod{pq}$.

15. (a) By Problem 21 of Section 2.3, $p \mid 2^p - 2$ implies that $2^p - 1 \mid 2^{2^p - 2} - 1$ or $M_p \mid 2^{M_p - 1} - 1$; hence $M_p \mid 2^{M_p} - 2$, which makes M_p a pseudoprime.

   (b) Since $2^{n+1} \mid 2^{2^n}$, Problem 21 of Section 2.2 gives $2^{2^{n+1}} - 1 \mid 2^{F_n - 1} - 1$. But $2^{2^{n+1}} - 1 = F_n(F_n + 2)$; hence $F_n \mid 2^{F_n - 1} - 1$ and so $F_n \mid 2^{F_n} - 2$.

17. Modulo 31,

$$11^{341} \equiv (11^{30})^{11} \cdot 11^{11} \equiv (11^3)^3 \cdot 11^2 \equiv (-2)^3(-3) \equiv 24 \equiv -7:$$

hence $11^{341} \not\equiv 11 \pmod{31}$. This implies that

$$11^{341} \not\equiv 11 \pmod{11 \cdot 31},$$

so that $341 = 11 \cdot 31$ is not an absolute pseudoprime.

19. Let $n = (6k+1)(12k+1)(18k+1)$, where each factor is prime. Since $6k$, $12k$ and $18k$ all divide $n - 1 = 36k(36k^2 + 11k + 1)$, the integer n is an absolute pseudoprime by Theorem 5.3.

21. Modulo 7, $1111 \equiv -2$ implies that

$$
\begin{aligned}
2222^{5555} + 5555^{2222} &\equiv (-4)^{5555} + 4^{2222} \\
&\equiv ((-4)^5)^{6 \cdot 185 + 1} + (4^2)^{6 \cdot 185 + 1} \\
&\equiv (-4)^5 + 4^2 \equiv 0.
\end{aligned}
$$

5.3   Wilson's Theorem

1. (a) $16 \equiv -1 \pmod{17}$ and so $15!16 \equiv 15!(-1) \equiv -1 \pmod{17}$. Thus $15! \equiv 1 \pmod{17}$.

    (b) $28! \equiv -1 \pmod{29}$ and so

    $$26! \cdot 27 \cdot 28 \equiv 26!(-2)(-1) \equiv -1 \pmod{29}.$$

    Thus $2(26!) \equiv -1 \equiv 28 \pmod{29}$.

5. (a) If $n > 1$ is prime, then $(n-1)! \equiv -1 \pmod{n}$, so that $(n-2)!(n-1) \equiv (n-2)!(-1) \equiv -1 \pmod{n}$; this yields $(n-2)! \equiv 1 \pmod{n}$. By the converse of Wilson's Theorem, n must be prime.

    (b) Suppose n is composite, say $n = ab$ where $0 < a \le b < n$. If $a \ne b$, then a and b are both factors of $(n-1)!$. Hence $ab = n$ is a factor of $(n-1)!$, so that $(n-1)! \equiv 0 \pmod{n}$. If $a = b$, so that $n = a^2$, then a and $2a$ are both factors of $(n-1)!$ provided that $2a \le n - 1 = a^2 - 1$; this holds for $a \ne 2$, or $n \ne 4$. Since a and $2a$ are both factors of $(n-1)!$ for $n \ne 4$, $a^2 = n$ divides $(n-1)!$; hence $(n-1)! \equiv 0 \pmod{n}$.

7. Using the theorems of Wilson and Fermat,

    $$\begin{aligned}
    a^p + (p-1)!a &\equiv a + (-1)a \equiv 0 \pmod{p}, \\
    (p-1)!a^p + a &\equiv (-1)a + a \equiv 0 \pmod{p}
    \end{aligned}$$

    for any integer a.

9. Modulo $p > 2$, we have

    $$\begin{aligned}
    2 &\equiv -(p-2), \quad 4 \equiv -(p-4), \quad 6 \equiv -(p-6), \ldots, \\
    p - 5 &\equiv -5, \qquad p - 3 \equiv -3, \qquad p - 1 \equiv -1.
    \end{aligned}$$

    The result of multiplying these congruences together is

    $$2 \cdot 4 \cdot 6 \cdots \equiv (-1)^{(p-1)/2} 1 \cdot 3 \cdot 5 \cdots (p-2).$$

    Using Wilson's Theorem, this leads to

    $$(-1)^{(p-1)/2} 1^2 \cdot 3^2 \cdot 5^2 \cdots (p-2)^2 \equiv (p-1)! \equiv -1.$$

    Now multiply both sides of this congruence by $(-1)^{(p-1)/2}$ to get

    $$1^2 \cdot 3^2 \cdot 5^2 \cdots (p-2)^2 \equiv (-1)^{(p+1)/2} \pmod{p}.$$

13. Assume $\sqrt{2} = a/b$, with $\gcd(a, b) = 1$. Then $a^2 = 2b^2$ and so $a^2 + b^2 = 3b^2$. This implies that $a^2 + b^2 \equiv 0 \pmod{3}$. By Problem 12, $a \equiv b \equiv 0 \pmod{3}$, whence $\gcd(a, b) \geq 3$ which is a contradiction.

15. Since $30! \equiv -1 \pmod{31}$, it follows that $29! \equiv 1 \pmod{31}$. Hence

$$4(29!) + 5! \equiv 4 \cdot 1 + 120 \equiv 124 \equiv 0 \pmod{31}.$$

17. Since $a^p \equiv a \pmod{p}$ and $a^q \equiv a \pmod{q}$,

$$
\begin{aligned}
a^{pq} - a^p - a^q + a &\equiv a^q - a - a^q + a \equiv 0 \pmod{p}, \\
a^{pq} - a^p - a^q + a &\equiv a^p - a^p - a + a \equiv 0 \pmod{q}.
\end{aligned}
$$

Hence, pq divides the above expression.

5.4   The Fermat-Kraitchik Factorization Method

1. (a) $2279 = 48^2 - 5^2 = (48 + 5)(48 - 5) = 53 \cdot 43$.

   (b) $10541 = 105^2 - 22^2 = (105 + 22)(105 - 22) = 127 \cdot 83$.

   (c) $340663 = 592^2 - 99^2 = (592 + 99)(592 - 99) = 691 \cdot 493$.

3. $2^{11} = 1 = 2047 = 56^2 - 33^2 = (56 + 33)(56 - 33) = 89 \cdot 23$.

5. (a) Since $138^2 \equiv 67^2 \pmod{2911}$,

$$
\begin{aligned}
\gcd(138 - 67, 2911) &= \gcd(71, 2911) = 71 \text{ and} \\
\gcd(138 + 67, 2911) &= \gcd(205, 2911) = 41
\end{aligned}
$$

   are divisors of 2911; in fact, $2911 = 71 \cdot 41$.

   (b) The congruence $177^2 \equiv 92^2 \pmod{4573}$ implies that

$$
\begin{aligned}
\gcd(177 - 92, 4573) &= \gcd(85, 4573) = 17 \text{ and} \\
\gcd(177 + 92, 4573) &= \gcd(269, 4573) = 269
\end{aligned}
$$

   are divisors of 4573; hence $4573 = 17 \cdot 269$.

   (c) From $208^2 \equiv 93^2 \pmod{6923}$, we get

$$
\begin{aligned}
\gcd(208 - 93, 6923) &= \gcd(115, 6923) = 23 \text{ and} \\
\gcd(208 + 93, 6923) &= \gcd(301, 6923) = 301
\end{aligned}
$$

   as the divisors of 6923; $6923 = 23 \cdot 301$.

7. (a) The relations $95^2 - 2 \cdot 4537 = -7^2$ and $213^2 - 10 \cdot 4537 = -1$ imply that $95^2 \equiv -7^2 \pmod{4537}$ and $213^2 \equiv -1 \pmod{4537}$. Thus, $(95 \cdot 213)^2 \equiv 7^2 \pmod{4537}$ or $2087^2 \equiv 7^2 \pmod{4537}$. This means that

$$\begin{aligned} \gcd(2087 - 7, 4537) &= \gcd(2080, 4537) = 13 \text{ and} \\ \gcd(2087 + 7, 4537) &= \gcd(2094, 4537) = 349 \end{aligned}$$

are divisors of 4537; that is, $4537 = 13 - 349$.

(b) The relations $120^2 - 14429 = -29$ and $3003^2 - 625 \cdot 14429 = -116$ imply that

$$120^2 \equiv -29 \pmod{14429} \text{ and } 3003^2 \equiv -116 \pmod{14429}.$$

Multiplying these congruences yields

$$(120 \cdot 3003)^2 \equiv (29 \cdot 2)^2 \pmod{14429}$$

or

$$14064^2 \equiv 58^2 \pmod{14429}.$$

Thus,

$$\begin{aligned} \gcd(14064 - 58, 14429) &= \gcd(14006, 14429) = 47 \text{ and} \\ \gcd(14064 + 58, 14429) &= \gcd(14122, 14429) = 307 \end{aligned}$$

are divisors of 14429; hence, $14429 = 47 \cdot 307$.

Chapter 6

Number-Theoretic Functions

6.1 The Sum and Number of Divisors

1. Let $m = p_1^{k_1} p_2^{k_2} \cdots p_r^{k_r}$ and $n = p_1^{j_1} p_2^{j_2} \cdots p_r^{j_r}$ where $k_i, j_i \geq 0$. Put $u_i = \min\{k_i, j_i\}$ and $v = \max\{k_i, j_i\}$ for $1 \leq i \leq r$. Suppose $d = p_1^{u_1} p_2^{u_2} \cdots p_r^{u_r}$. Since $u_i \leq k_i$, Theorem 6-1 implies that $d|m$; likewise $u_i \leq j_i$ yields $d|n$. Thus d is a common divisor of m and n.

   Let d' be any common divisor of m and n. Then $d' = p_1^{h_1} p_2^{h_2} \cdots p_r^{h_r}$ with $h_i \geq 0$. Since $d'|m$, $h_i \leq k_i$, and since $d'|n$, $h_i \leq j_i$; hence $h_i \leq u_i$. This means that $d'|d$ by Theorem 6-1, and so $d' \leq d$. It follows that $d = \gcd(m, n)$.

   A similar argument gives $\operatorname{lcm}(m, n) = p_1^{v_1} p_2^{v_2} \cdots p_r^{v_r}$.

3. In the notation of Problem 1,

$$
\begin{aligned}
\gcd(m, n)\operatorname{lcm}(m, n) &= (p_1^{u_1} p_2^{u_2} \cdot p_r^{u_r})(p_1^{v_1} p_2^{v_2} \cdots p_r^{v_r}) \\
&= p_1^{u_1+v_1} p_2^{u_2+v_2} \cdots p_r^{u_r+v_r} \\
&= p_1^{k_1+j_1} p_2^{k_2+j_2} \cdots p_r^{k_r+j_r} \\
&= (p_1^{k_1} p_2^{k_2} \cdots p_r^{k_r})(p_1^{j_1} p_2^{j_2} \cdots p_r^{j_r}) = mn
\end{aligned}
$$

5. (a)

$$
\begin{array}{ll}
\tau(3655) = \tau(5 \cdot 7 \cdot 43) = 8, & \tau(4503) = \tau(3 \cdot 19 \cdot 73) = 8, \\
\tau(3656) = \tau(2^3 \cdot 457) = 8, & \tau(4505) = \tau(2^3 \cdot 563) = 8, \\
\tau(3657) = \tau(3 \cdot 23 \cdot 53) = 8, & \tau(4505) = \tau(5 \cdot 17 \cdot 53) = 8, \\
\tau(3658) = \tau(2 \cdot 31 \cdot 59) = 8, & \tau(4506) = \tau(2 \cdot 3 \cdot 751) = 8.
\end{array}
$$

(b)

$$\begin{aligned}
\sigma(14) &= \sigma(2 \cdot 7) = 24 = \sigma(3 \cdot 5) = \sigma(15). \\
\sigma(206) &= \sigma(2 \cdot 103) = 312 = \sigma(3^2 \cdot 23) = \sigma(207). \\
\sigma(957) &= \sigma(3 \cdot 11 \cdot 29) = 1440 = \sigma(2 \cdot 479) = \sigma(958).
\end{aligned}$$

7. (a) The result is trivial when $n = l$, so assume $n > 1$. If $n = a^2$, where $a = p_1^{k_1} p_2^{k_2} \cdots p_r^{k_r}$, then $n = p_1^{2k_1} p_2^{2k_2} \cdots p_r^{2k_r}$; hence

$$\tau(n) = (2k_1 + 1)(2k_2 + 1) \cdots (2k_r + 1),$$

which is odd. Conversely, suppose that $n = p_1^{k_1} p_2^{k_2} \cdots p_r^{k_r}$ and that $\tau(n) = (k_1 + 1)(k_2 + 1) \cdots (k_r + 1)$ is an odd integer; then $k_i + 1$ is odd for each i, so that $k_i = 2j_i$ for some j_i. Thus

$$n = p_1^{2j_1} p_2^{2j_2} \cdots p_r^{2j_r} = (p_1^{j_1} p_2^{j_2} \cdots p_r^{j_r})^2 = a^2.$$

(b) Suppose that $\tau(n)$ is odd and $n = 2^k m$, where $m \geq 1$ is odd and $k \geq 0$. Since $\sigma(n) = (2^{k+1} - 1)\sigma(m)$ is odd, $\sigma(m)$ and its divisors will be odd. Assuming $m > 1$, let $m = p_1^{k_1} p_2^{k_2} \cdots p_r^{k_r}$, with $p_i \neq 2$. Then

$$\begin{aligned}
\sigma(m) &= \sigma(p_1^{k_1})\sigma(p_2^{k_2}) \cdots \sigma(p_r^{k_r}), \text{ where} \\
\sigma(p_i^{k_i}) &= 1 + p_i + \cdots + p_i^{k_i}.
\end{aligned}$$

For $\sigma(p_i)$ to be odd, k_i must be even; say $k_i = 2j_i$. This gives $m = p_1^{2j_1} p_2^{2j_2} \cdots p_r^{2j_r} = b^2$. If $k = 2j$ is even, then $n = (2^j b)^2 = a^2$, while if $k = 2j + 1$ is odd, then $n = 2(2^j b)^2 = 2a^2$.

On the other hand, suppose that $n = a^2$ or $n = 2a^2$. Then we can write $n = 2^k p_1^{2k_1} \cdots p_r^{2k_r}$, where $p_i \neq 2$ and $k \geq 0$. Hence

$$\sigma(n) = (2^{k+1} - 1)\sigma(p_1^{2k_1}) \cdots \sigma(p_r^{2k_r}).$$

Since $\sigma(p_i^{2k_i}) = 1 + p_i + p_i^2 + \cdots + p_i^{2k_i}$ is odd for each i, $\sigma(n)$ must be odd.

9. Suppose n is square-free, say $n = p_1 p_2 \cdots p_r$, for distinct primes p_i. Then $\sigma(n) = (1 + 1)(1 + 1) \cdots (1 + 1) = 2^r$.

11. For $k > 1$, $\sigma(n) = k$ is satisfied by $n = p^{k-1}$ for any prime p. By Problem 10(a), the equation $\sigma(n) = k$ implies that $n < k$: hence $\sigma(n) = k$ has only a finite number of solutions, if any.

13. Let $m = 5k$ and $n = 4k$, where $\gcd(k, 10) = 1$. Then
$$\begin{aligned} \sigma(m^2) &= \sigma(5^2)\sigma(k^2) = 31\sigma(k^2), \quad \text{and} \\ \sigma(n^2) &= \sigma(2^4)\sigma(k^2) = 31\sigma(k^2). \end{aligned}$$

15. If n and $n+2$ are twin primes, then
$$\sigma(n+2) = 1 + (n+2) = (1+n) + 2 = \sigma(n) + 2.$$
Also,
$$\begin{aligned} \sigma(434) + 2 &= \sigma(2 \cdot 7 \cdot 31) + 2 = 770 = \sigma(2^2 \cdot 109) = \sigma(436), \\ \sigma(8575) + 2 &= \sigma(5^2 \cdot 7^3) + 2 = 12402 = \sigma(3^2 \cdot 953) = \sigma(8577). \end{aligned}$$

17. Since $f(mn) = (mn)^k = m^k n^k = f(m)f(n)$, f is (completely) multiplicative.

19. Suppose f and g are multiplicative functions: and $\gcd(m, n) = 1$. Then
$$\begin{aligned} (fg)(mn) &= f(mn)g(mn) = f(m)f(n)g(m)g(n) \\ &= f(m)g(m)f(n)g(n) \\ &= (fg)(m)(fg)(n), \end{aligned}$$
whence fg is also multiplicative.

21. By Problem 19 and Theorem 6-4, the functions
$$F(n) = \sum_{d|n} \tau(d)^3 \quad \text{and} \quad G(n) = \left(\sum_{d|n} \tau(d)\right)^2$$
are both multiplicative. Thus to see that $F = G$, it is enough to show that they agree on powers of primes. Now
$$\begin{aligned} F(p^k) &= \tau(1)^3 + \tau(p)^3 + \tau(p^2)^3 + \cdots \tau(p^k)^3 \\ &= 1^3 + 2^3 + 3^3 + \cdots + (k+1)^3 = \left[\frac{(k+1)(k+2)}{2}\right]^3, \end{aligned}$$
by Problem 1(e) of Section 1.1. Also
$$\begin{aligned} G(p^k) &= (\tau(1) + \tau(p) + \tau(p^2) + \cdots + \tau(p^k))^2 \\ &= (1 + 2 + 3 + \cdots + (k+1))^2 = \left[\frac{(k+1)(k+2)}{2}\right]^2, \end{aligned}$$
whence $F(p^k) = G(p^k)$.

23. (a) By Problem 19 and Theorem 6-4, the functions

$$F(n) = \sum_{d|m} \sigma(n) \quad \text{and} \quad G(n) = \sum_{d|n} (n/d)\tau(n)$$

are multiplicative. Hence, it suffices to show that they agree on powers of a prime. Now

$$
\begin{aligned}
F(p^k) &= \sigma(1) + \sigma(p) + \sigma(p^2) + \cdots + \sigma(p^k) \\
&= 1 + (1+p) + (1+p+p^2) + \cdots + (1+p+p^2+\cdots+p^k) \\
&= (k+1) + kp + (k-1)p^2 + \cdots + 2p^{k-1} + p^k,
\end{aligned}
$$

and

$$
\begin{aligned}
G(p^k) &= \frac{p^k}{1\tau(1)} + \frac{p^k}{p\tau(p)} + \frac{p^k}{p^2\tau(p^2)} + \cdots + \frac{p^k}{p^k\tau(p^k)} \\
&= p^k + 2p^{k-1} + 3p^{k-2} + \cdots + (k+1)
\end{aligned}
$$

so that $F(p^k) = G(p^k)$.

(b) Since they are multiplicative functions, it is enough to show that

$$H(n) = \sum_{d|n} (n/d)\sigma(d) \quad \text{and} \quad J(n) = \sum_{d|n} d\tau(d)$$

satisfy $H(p^k) = J(p^k)$ for any prime p. Now

$$
\begin{aligned}
H(p^k) &= \frac{p^k}{1\sigma(1)} + \frac{p^k}{p\sigma(p)} + \frac{p^k}{p^2\sigma(p^2)} + \cdots + \frac{p^k}{p^k\sigma(p^k)} \\
&= p^k + p^{k-1}(1+p) + p^{k-2}(1+p+p^2) \\
&\qquad + \cdots + (1+p+p^2+\cdots+p^k) \\
&= (k+1)p^k + kp^{k-1} + (k-1)p^{k-2} + \cdots + 2p + 1,
\end{aligned}
$$

and

$$
\begin{aligned}
J(p^k) &= \tau(1) + p\tau(p) + p^2\tau(p^2) + \cdots + p^k\tau(p^k) \\
&= 1 + 2p + 3p^2 + \cdots + (k+1)p^k.
\end{aligned}
$$

6.2   The Möbius Inversion Formula

1. (a) Of any four consecutive integers $n, n+1, n+2, n+3$, one will be divisible by 4 and hence have μ-value of 0.

(b) For $n \geq 3$,

$$\sum_{k=1}^{n} \mu(k!) = \mu(1!) + \mu(2!) + \mu(3!) + \cdots + \mu(n!)$$
$$= \mu(1) + \mu(2) + \mu(6) = 1 + (-1) + 1 = 1.$$

3. If f is a multiplicative function, so is μf; hence, the function F defined by $F(n) = \sum \mu(d) f(d)$ is also multiplicative. For a prime p,

$$F(p^k) = \mu(1)f(1) + \mu(p)f(p) + \mu(p^2)f(p^2) + \cdots + \mu(p^k)f(p^k)$$
$$= \mu(1)f(1) + \mu(p)f(p) = 1 - f(p).$$

If $n = p_1^{k_1} p_2^{k_2} \cdots p_r^{k_r}$, it follows that

$$F(n) = F(p_1^{k_1}) F(p_2^{k_2}) \cdots F(p_r^{k_r})$$
$$= (1 - f(p_1))(1 - f(p_2)) \cdots (1 - f(p_r)).$$

5. Let $S(n)$ denote the number of square-free divisors of n. If $n > 1$ and $n = p_1^{k_1} p_2^{k_2} \cdots p_r^{k_r}$, then a divisor d of n will be square-free provided that $d = p_1^{j_1} p_2^{j_2} \cdots p_r^{j_r}$, where each j_i is either 0 or 1. There are 2^r such divisors, hence $S(n) = 2^r$.

Now the function $|\mu|$ is multiplicative. For if $\gcd(m, n) = 1$, then

$$|\mu|(mn) = |\mu(mn)| = |\mu(m)\mu(n)| = |\mu(m)||\mu(n)|$$
$$= |\mu|(m) \cdot |\mu|(n).$$

Thus, the function F defined by $F(n) = \sum_{d|n} |\mu(d)|$ is also multiplicative. For, with p a prime,

$$F(p^k) = |\mu(1)| + |\mu(p)| + |\mu(p^2)| + \cdots + |\mu(p^k)|$$
$$= |\mu(1)| + |\mu(p)| = 1 + |-1| = 2,$$

and in consequence

$$F(n) = F(p_1^{k_1}) F(p_2^{k_2}) \cdots F(p_r^{k_r}) = 2 \cdot 2 \cdots 2 = 2^r = S(n).$$

7. (a) Let $m = p_1^{k_1} p_2^{k_2} \cdots p_r^{k_r}$ and $n = q_1^{j_1} q_2^{j_2} \cdots q_s^{j_s}$, with $p_i \neq q_j$, so that $\gcd(m, n) = 1$. Then

$$\lambda(mn) = (-1)^{k_1 + \cdots + k_r + j_1 + \cdots + j_s}$$
$$= (-1)^{k_1 + \cdots + k_r} \cdot (-1)^{j_1 + \cdots + j_s} = \lambda(m)\lambda(n).$$

(b) By Theorem 6-4, the function F defined by $F(n) = \sum_{d|n} \lambda(d)$ is multiplicative. For a prime p,

$$\begin{aligned} F(p^k) &= \lambda(1) + \lambda(p) + \lambda(p^2) + \cdots + \lambda(p^k) \\ &= 1 + (-1) + (-1)^2 + \cdots + (-1)^k \\ &= \begin{cases} 1 & \text{if } k = 2j; \\ 0 & \text{if } k = 2j+1 \end{cases} \end{aligned}$$

Thus $F(n) = F(p_1^{k_1})F(p_2^{k_2})\cdots F(p_r^{k_r}) = 1$ if $k_i = 2j_i$ for $1 \leq i \leq$ r (that is, $n = m^2$), and 0 otherwise.

6.3   The Greatest Integer Function

1. Given a and $b > 0$, we have $a = qb + r$ for integers q and $0 \leq r < b$. Thus $a/b = q + r/b$, with $r/b < 1$: hence $[a/b] = q$, so $a = [a/b]q + r$.

3. The highest power of 5 dividing 1000! is

$$[1000/5] + [1000/25] + [1000/125] + [1000/625] = 200 + 40 + 8 + 1 = 249,$$

while the highest power of 7 dividing 2000! is

$$[2000/7] + [2000/49] + [2000/343] = 285 + 40 + 5 = 330.$$

5. (a) The number of zeros with which 1000! terminates is equal to the exponent of 5 in the prime factorization of 1000!, namely, 249.

   (b) In order that $[n/5] = 37$, we begin by considering $n \leq 185$. It is easily checked that for $n \leq 149$,

   $$[n/5] + [n/25] + [n/125] + \cdots \leq 36,$$

   while for $155 \leq n$,

   $$[n/5] + [n/25] + [n/125] + ??? \geq 38.$$

   Thus n must lie in the range $150 \leq n \leq 154$.

7. Let $n = a_k p^k + \cdots + a_2 p^2 + a_1 p + a_0$, where $0 \leq a_i < p$. Then

$$\begin{aligned} [n/p] &= a_k p^{k-1} + \cdots + a_2 p + a_1 \\ \left[n/p^2\right] &= a_k p^{k-2} + \cdots + a_2 \\ &\vdots \\ \left[n/p^k\right] &= a_k \end{aligned}$$

which implies that

$$
\begin{aligned}
a_0 + p[n/p] &= n \\
a_1 + p[n/p^2] &= [n/p] \\
a_2 + p[n/p^2] &= [n/p^2] \\
&\vdots \\
a_{k-1} + p[n/p^k] &= [n/p^{k-1}] \\
a_k &= [n/p^k]
\end{aligned}
$$

Adding these k equations leads to

$$
(a_0 + a_1 + a_2 + \cdots + a_k) + p \sum_{i=1}^{k} [n/p^i] = n + \sum_{i=1}^{k} [n/p^i]
$$

or $(p-1) \sum_{i=1}^{k} [n/p^i] = n - (a_0 + a_1 + a_2 + \cdots + a_k)$.

9. If $n = a_k 5^k + \cdots + a_2 5^2 + a_1 5 + a_0$, then it is desired that

$$
\frac{n - (a_k + \cdots + a_2 + a_1 + a_0)}{4} = 100.
$$

Since $a_k + \cdots + a_2 + a_1 + a_0 \geq 1$, we begin by considering n with $(n-1)/4 = 100$. Now $405 - (3 + 1 + 1 + 0) = 400$, where $405 = 3 \cdot 5^3 + 1 \cdot 5^2 + 1 \cdot 5 + 0$, and

$$
[405/5] + [405/25] + [405/125] = 81 + 16 + 3 = 100.
$$

Hence 405 is one solution, as is $406 = 3 \cdot 5^3 + 1 \cdot 5^2 + 1 \cdot 5 + 1$.

13. For $N \geq 1$, Corollary 1 to Theorem 6-11 gives

(a)

$$
\sum_{n=1}^{N} \tau(n) - \sum_{n=1}^{N} [2N/n] = \sum_{n=1}^{2N} [2N/n] - \sum_{n=1}^{N} [2N/n]
$$

$$
= \sum_{n=N+1}^{2N} [2N/n] = 1 + 1 + \cdots + 1 = N;
$$

and

(b) $\sum_{n=1}^{N} ([N/n] - [(N-1)/n]) = \sum_{n=1}^{N} \tau(n) - \sum_{n=1}^{N-1} \tau(n) = \tau(N)$.

6.4   An Application to the Calendar

1. (a) The number of leap years between 1600 (exclusive) and 1825 (inclusive) is

$$L_{1825} = [1825/4] - [1825/100] + [1825/400] - 388$$
$$= 456 - 18 + 4 - 388 = 54.$$

   (b) The number of leap years between 1600 (exclusive) and 1950 (inclusive) is

$$L_{1950} = [1950/4] - [1950/1001 + [1950/400] - 388$$
$$= 487 - 19 + 4 - 388 = 84.$$

   (c) The number of leap years between 1600 (exclusive) and 2075 (inclusive) is

$$L_{2075} = [2075/4] - [2075/100] + [2075/400] - 388$$
$$= 518 - 20 + 5 - 388 = 115.$$

3. (a) For November 19, 1863 in the Gregorian calendar, $d = 19$, $m = 9$, $c = 18$ and $y = 63$. This implies that

$$w = 19 + [(2.6)9 - 0.2] - 2 \cdot 18 + 63 + [18/4] + [63/4] \equiv 4 \pmod 7.$$

   Hence, November 19, 1863 was a Thursday.

   (b) For April 18, 1906, we have $d = 18$, $m = 2$, $c = 19$, and $y = 6$:
   Thus

$$w = 18 + [(2.6)2 - 0.2] - 2 \cdot 19 + 6 + [19/4] + [6/4] \equiv 3 \pmod 7$$

   which means that April 18, 1906 fell on a Wednesday.

   (c) For November 11, 1918, it follows that $d = 11$, $m = 9$, $c = 19$, $y = l8$ and so

$$w = 11 + [(2.6)9 - 0.2] - 2 \cdot 19 + 18 + [19/4] + [18/4] \equiv 1 \pmod 7.$$

   Thus, November 11, 1918 was a Monday.

   (d) For October 24, 1929, we have $d = 24$, $m = 8$, $c = 19$, and $y = 29$. This implies that

$$w = 24 + [(2.6)8 - 0.2] - 2 \cdot 19 + 29 + [19/4] + [29/4] \equiv 4 \pmod 7.$$

   Hence, October 24, 1929 fell on a Thursday.

(e) For June 6, 1944, we have $d = 6$, $m = 4$, $c = 19$, and $y = 44$. Thus

$$w = 6 + [(2.6)4 - 0.2] - 2 \cdot 19 + 44 + [19/4] + [44/4] \equiv 2 \pmod 7$$

so that June 6, 1944 was a Tuesday.

(f) For February 15, 1898, we have $d = 15$, $m = 12$, $c = 18$, and $y = 97$. Hence

$$w = 15 + [(2.6)12 - 0.2] - 2 \cdot 18 + 97 + [18/4] + [97/4] \equiv 2 \pmod 7$$

making February 15, 1898 a Tuesday.

5. (a) The Mondays in March 2010 satisfy the congruence

$$1 \equiv d + [2.6 - 0.2] - 2 \cdot 20 + 10 + [20/4] + [10/4] \pmod 7.$$

Since this reduces to $d \equiv 1 \pmod 7$, the dates for Mondays are 1, 8, 15, 22 and 29.

(b) For Friday the 13th in the year 2010, we have

$$5 \equiv 13 + [(2.6)m - 0.2] - 2 \cdot 20 + 10 + [20/4] + [10/4] \pmod 7.$$

That is, $[(2.6)m - 0.2] \equiv 1 \pmod 7$ with $1 \leq m < 10$. This congruence holds only for $m = 6$. Hence, August is the only month from March to December in which the 13th is on a Friday. January and February are excluded by considering the congruence

$$5 \equiv 13 + [(2.6)m - 0.2] - 2 \cdot 20 + 9 + [20/4] + [9/4] \pmod 7.$$

where $m = 11$ or 12.

Chapter 7

Euler's Generalization of Fermat's Theorem

7.2 Euler's Phi-Function

1.

$$
\begin{aligned}
\phi(1001) &= \phi(7 \cdot 11 \cdot 13) = \phi(7)\phi(11)\phi(13) = 6 \cdot 10 \cdot 12 = 720. \\
\phi(5040) &= \phi(2^4 \cdot 3^2 \cdot 5 \cdot 7) \\
&= \phi(2^4)\phi(3^2)\phi(5)\phi(7) = 8 \cdot 6 \cdot 4 \cdot 6 = 1152. \\
\phi(36000) &= \phi(2^5)\phi(3^2)\phi(5^3) = 16 \cdot 6 \cdot 120 = 9600.
\end{aligned}
$$

3. Let $m = 3^k \cdot 568 = 3^k \cdot 2^3 \cdot 71$ and
$n = 3^k \cdot 638 = 3^k \cdot 2 \cdot 11 \cdot 29$ for $k \geq 1$. Then

$$
\begin{aligned}
\tau(m) &= \tau(3^k)\tau(2^3)\tau(71) = (k+1) \cdot 4 \cdot 2 = 8(k+1) \text{ and} \\
\tau(n) &= \tau(3^k)\tau(2)\tau(11)\tau(29) = (k+1) \cdot 2 \cdot 2 \cdot 2 = 8(k+1).
\end{aligned}
$$

Thus $\tau(m) = \tau(n)$. Also,

$$
\begin{aligned}
\sigma(m) &= \sigma(3^k\sigma(2^3\sigma(71) = \frac{3^{k+1}-1}{2} \cdot 15 \cdot 72 \\
&= 540(3^{k+1}-1) \text{ and} \\
\sigma(n) &= \sigma(3^k)\sigma(2)\sigma(29) = \frac{3^{k+1}-1}{2} \cdot 3 \cdot 12 \cdot 30 \\
&= 540(3^{k+1}-1);
\end{aligned}
$$

hence $\sigma(m) = \sigma(n)$. Finally,

$$
\begin{aligned}
\phi(m) &= \phi(3^k)\phi(2^3)\phi(71) = (3^k - 3^{k-1}) \cdot 4 \cdot 70 \\
&= 280(3^k - 3^{k-1}) \text{ and} \\
\phi(n) &= \phi(3^k)\phi(2)\phi(11)\phi(29) = (3^k - 3^{k-1}) \cdot 1 \cdot 10 \cdot 28 \\
&= 280(3^k - 3^{k-1}),
\end{aligned}
$$

so that $\phi(m) = \phi(n)$.

5. Suppose that $n = 2(2p - 1)$, where p and $2p - 1$ are both odd primes. Then

$$
\begin{aligned}
\phi(n) &= \phi(2)\phi(2p - 1) = 1 \cdot (2p - 2) \\
&= 2p - 2, \text{ and} \\
\phi(n + 2) &= \phi(4p) = \phi(4)\phi(p) \\
&= 2 \cdot (p - 1) = 2p - 2.
\end{aligned}
$$

Thus, $\phi(n) = \phi(n + 2)$.

7. (a) Assume that n has the prime factorization $n = 2^{k_0}p_1^{k_1}\cdots p_r^{k_r}$, where $k_0 \geq 1$ and $k_i \geq 1$ for $i = 1, 2, ..., r$. Since $p - 1 > \sqrt{p}$ for any odd prime and $k - \frac{1}{2} = \frac{k}{2} + \frac{k-1}{2} \geq \frac{k}{2}$ for $k \geq 1$,

$$
\begin{aligned}
\phi(n) &= 2^{k_0-1}p_1^{k_1-1}\cdots p_r^{k_r-1}(p_1 - 1)\cdots(p_r - 1) \\
&> 2^{k_0-1}p_1^{k_1-1}\cdots p_r^{k_r-1}\sqrt{p_1}\cdots\sqrt{p_r} \\
&= \left(2^{k_0}p_1^{k_1-(1/2)}\cdots p_r^{k_r-(1/2)}\right)/2 \\
&\geq \left(2^{k_0/2}p_1^{k_1/2}\cdots p_r^{k_r/2}\right) = \frac{\sqrt{n}}{2}.
\end{aligned}
$$

If $k_0 = 0$, so that $n = p_1^{k_1}\cdots p_r^{k_r}$, then $\phi(n) > \sqrt{n} > \sqrt{n}/2$; if $n = 2^{k_0}$ for $k_0 \geq 1$, then $\phi(n) = 2^{k_0-1} = (2^{k_0})/2 = n/2 \geq \sqrt{n}/2$. In any event, $\phi(n) \geq \sqrt{n}/2$. If $n > 1$ has the prime factorization $n = q_1^{k_1}q_2^{k_2}\cdots q_s^{k_s}$, then

$$
\phi(n) = n(1 - 1/q_1)(1 - 1/q_2)\cdots(1 - 1/q_s) < n
$$

since $1 - 1/q_i < 1$ for each i. Thus, for any $n \geq 1$,

$$
\sqrt{n}/2 \leq \phi(n) \leq n.
$$

(b) Let the integer $n > 1$ have r distinct prime factors, say $n = p_1^{k_1} p_2^{k_2} \cdots p_r^{k_r}$. Then, since $1 - 1/p_i \geq 1/2$ for all i,

$$\phi(n) = n(1 - 1/p_1)(1 - 1/p_2) \cdots (1 - 1/p_r) \geq n/2^r.$$

(c) Suppose that $n > 1$ is composite with smallest prime factor p. Then $p \leq \sqrt{n}$, so that $\sqrt{n} \leq n/p$. Hence, if $n = p^k p_1^{k_1} \cdots p_r^{k_r}$, it follows that

$$
\begin{aligned}
\phi(n) &= n(1 - 1/p)(1 - 1/p_1) \cdots (1 - 1/p_r) \\
&\leq n(1 - 1/p) = n - n/p \leq n - \sqrt{n}
\end{aligned}
$$

9. (a) Assume that n and $n + 2$ are both primes. Then

$$\phi(n + 2) = n + 1 = (n - 1) + 2 = \phi(n) + 2.$$

(b) Let p and $2p + 1$ both be odd primes. Then, for $n = 4p$,

$$
\begin{aligned}
\phi(n + 2) &= \phi(4p + 2) = \phi(2)\phi(2p + 1) \\
&= 1 \cdot (2p) = 2(p - 1) + 2 = \phi(4)\phi(p) + 2 \\
&= \phi(n) + 2.
\end{aligned}
$$

11. (a) Let $n > 1$ be such that $\phi(n) | n - 1$, and assume to the contrary that n is not square-free. Then $n = p_1^{k_1} p_2^{k_2} \cdots p_r^{k_r}$ for primes $p_1, p_2, \ldots, p_r$ and some exponent $k_i \geq 2$; for simplicity, let $k_1 \geq 2$. Thus,

$$\phi(p_1^{k_1}) = p_1^{k_1} - p_1^{k_1 - 1} = p_1 \cdot p_1^{k_1 - 2}(p_1 - 1),$$

so that $p_1 | \phi(p_1^{k_1})$. Since $\phi(n) = \phi(p_1^{k_1})\phi(p_2^{k_2} \cdots p_r^{k_r})$, it follows that $p_1 | \phi(n)$ and so $p_1 | n - 1$. Therefore, $p_1 | n$ and $p_1 | n - 1$ leading to the contradiction that $p_1 | 1$. This contradiction means that n is square-free.

(b) Let $n = 2k$ where $k \geq 1$. Then $\phi(n) = 2^{k-1} = n/2$ or $2\phi(n) = n$: hence, $\phi(n) | n$. If $n = 2^k 3^j$ for $k, j \geq 1$, then

$$\phi(n) = n \left(1 - \frac{1}{2}\right)\left(1 - \frac{1}{3}\right) = \frac{n}{3}.$$

Thus $3\phi(n) = n$ and so $\phi(n) | n$.

13. Assume that $d|n$ and that $n > 1, d > 1$ (the result being trivial otherwise). Let $n = p_1^{k_1} p_2^{k_2} \cdots p_r^{k_r}$, where $k_i \geq 1$ for $i = 1, 2, \ldots, r$. There is no harm in supposing that the first s primes in n appear in d, with $s \leq r$: then $d = p_1^{j_1} p_2^{j_2} \cdots p_s^{j_s}$ with $j_i \leq k_i$ for $i = 1, 2, \ldots, s$. Now

$$\phi(d) = \left(p_1^{j_1} - p_1^{j_1-1}\right) \cdots \left(p_s^{j_s} - p_s^{j_s-1}\right) \text{ and}$$

$$\begin{aligned}
\phi(n) &= \left(p_1^{k_1} - p_1^{k_1-1}\right) \cdots \left(p_s^{k_s} - p_s^{k_s-1}\right) \cdots \left(p_r^{k_r} - p_r^{k_r-1}\right) \\
&= p_1^{k_1-j_1}\left(p_1^{j_1} - p_1^{j_1-1}\right) \cdots p_s^{k_s-j_s}\left(p_s^{j_s} - p_s^{j_s-1}\right) \cdots \left(p_r^{k_r} - p_r^{k_r-1}\right) \\
&= \phi(d)\left[p_1^{k_1-j_1} \cdots p_s^{k_s-1} \cdots \left(p_r^{k_r} - p_r^{k_r-1}\right)\right]
\end{aligned}$$

Thus $\phi(d)|\phi(n)$.

15. (a) Let $n = 2^k 3^j$ for positive integers k and j. Then

$$\phi(n) = n\left(1 - \frac{1}{2}\right)\left(1 - \frac{1}{3}\right) = n \cdot \frac{1}{2} \cdot \frac{2}{3} = \frac{n}{3},$$

whence there are infinitely many integers n satisfying $\phi(n) = n/3$.

(b) Assume that there is an integer n for which $\phi(n) = n/4$. Thus $4|n$. This means that n has the prime factorization $n = 2^k p_1^{k_1} p_2^{k_2} \cdots p_r^{k_r}$, where $k \geq 2$, p_i is an odd prime and $k \geq 1$ for $i = 1, 2, \ldots r$. (Recall that, by Problem 4(e), $\phi(n) = n/2$ if and only if $n = 2k$ for $k \geq 1$.) Now

$$n/4 = \phi(n) = n\left(1 - \frac{1}{2}\right)\left(1 - \frac{1}{p_1}\right)\left(1 - \frac{1}{p_2}\right) \cdots \left(1 - \frac{1}{p_r}\right)$$

or, upon simplifying,

$$2(p_1 - 1)(p_2 - 1) \cdots (p_r - 1) = p_1 p_2 \cdots p_r.$$

Hence, $2|p_1 p_2 \cdots p_r$ or $2|p_i$ for some i, which is impossible. Thus, there is no n satisfying $\phi(n) = n/4$.

17. (a) Assume the integer $k > 0$. By Problem 7(a), $\sqrt{n}/2 \leq \phi(n)$ for any positive integer n. Thus, when $n > 4k^2$,

$$\phi(n) \geq \frac{\sqrt{4k^2}}{2} = k.$$

Since only a finite number of positive integers do not exceed $4k^2$, the equation $\phi(n) = k$ can have at most a finite number of solutions.

(b) For $k > 0$, suppose that the equation $\phi(n) = k$ has a unique solution, say $n = n_0$. If n_0 were odd, then by Problem 4(a), $\phi(2n_0) = \phi(n_0) = k$, contradicting the uniqueness of n_0: hence n_0 is even, or $n_0 = 2m$ for some m. Again, if m were odd, then

$$k = \phi(n_0) = \phi(2m) = \phi(m),$$

which would make m a solution to $\phi(n) = k$. Thus, m must be even, and so $4 | n_0$.

19. (a) Consider the equation $\phi(n) = 2p$, where p is a prime and $2p+1$ is composite. Assume the equation has a solution $n_0 = 2^k p_1^{k_1} \cdots p_r^{k_r}$, with p_i an odd prime. Since

$$2p = \phi(n_0) = 2^{k-1} p_1^{k_1-1} \cdots p_r^{k_r-1}(p_1 - 1) \cdots (p_r - 1),$$

with $p_i - 1$ even, n_0 can have at most one odd prime in its factorization: that is, $n_0 = 2^k q^j$, where $k \geq 0$, $j \geq 0$ and q is an odd prime. If $j > 2$, then

$$2p = \phi(n_0) = 2^{k-1} q^2 q^{j-3}(q - 1)$$

and so $q^2 | p$, which is impossible. On the other hand, if $k \geq 2$, then the equation

$$2p = \phi(n_0) = 2^{k-1} q^{j-1}(q - 1)$$

implies that $2^2 | 2p$, whence $p = 2$: this makes $2p + 1 = 5$, which is not composite. Thus $n_0 = 1$, q, q^2, $2q$ or $2q^2$: but none of these values provides a solution to $\phi(n) = 2p$, where $2p + 1$ is composite. The equation is not solvable.

(b) Since 7 is prime and $2 \cdot 7 + 1 = 15$ is composite, the equation $\phi(n) = 2 \cdot 7 = 14$ has no solution by part (a). However, $\phi(n) = 2k$ is solvable for $1 \leq k \leq 6$: $\phi(3) = 2$, $\phi(5) = 4$, $\phi(7) = 6$, $\phi(16) = 8$, $\phi(11) = 10$ and $\phi(13) = 12$.

21. For $n = 63457 = 23 \cdot 31 \cdot 89$, we have

$$
\begin{aligned}
\phi(n) &= \phi(23)\phi(31)\phi(89) \\
&= 22 \cdot 30 \cdot 88 = 2^5 \cdot 3 \cdot 5 \cdot 11^2 \\
\sigma(n) &= \sigma(23)\sigma(31)\sigma(89) \\
&= 24 \cdot 32 \cdot 90 = 2^9 \cdot 3^3 \cdot 5
\end{aligned}
$$

It follows that

$$\phi(n)\sigma(n) = (2^7 \cdot 3^2 \cdot 5 \cdot 11)^2 = 63360^2.$$

7.3   Euler's Theorem

1. (a) Let $\gcd(a, 1729) = 1$, where $1729 = 7 \cdot 13 \cdot 19$. Then $\gcd(a, 19) = 1$, so that $a^{\phi(19)} \equiv 1 \pmod{19}$ or $a^{18} \equiv 1 \pmod{19}$: thus, $a^{36} \equiv 1 \pmod{19}$.

    Also, $\gcd(a, 13) = 1$ and so $a^{\phi(13)} \equiv 1 \pmod{13}$ or $a^{12} \equiv 1 \pmod{13}$; as a result, $a^{36} \equiv 1 \pmod{13}$.

    Finally, $\gcd(a, 7) = 1$ implies that $a^{\phi(7)} \equiv 1 \pmod{7}$ or $a^{6} \equiv 1 \pmod{7}$; hence, $a^{36} \equiv 1 \pmod{7}$. It follows that $a^{37} \equiv a \pmod{7}$, $a^{37} \equiv a \pmod{13}$, and $a^{37} \equiv a \pmod{19}$; these are also valid if $7|a$, $13|a$ or $19|a$.

    Accordingly, $a^{37} \equiv a \pmod{7 \cdot 13 \cdot 19}$ for any a.

   (b) Suppose that $\gcd(a, 2370) = 1$, where $2370 = 2 \cdot 3 \cdot 5 \cdot 7 \cdot 13$. Since

   $$\phi(2) = 1, \ \phi(3) = 2, \ \phi(5) = 4, \ \phi(7) = 6 \ \text{ and } \ \phi(13) = 12,$$

   Euler's Theorem implies that

   $$\begin{aligned}
   a &\equiv 1 \pmod{2}, \\
   a^2 &\equiv 1 \pmod{2}, \\
   a^4 &\equiv 1 \pmod{5}, \\
   a^6 &\equiv 1 \pmod{7}, \text{ and} \\
   a^{12} &\equiv 1 \pmod{13}.
   \end{aligned}$$

   These congruences lead to

   $$\begin{aligned}
   a^{12} &\equiv 1 \pmod{2}, \\
   a^{12} &\equiv 1 \pmod{3}, \\
   a^{12} &\equiv 1 \pmod{5}, \text{ and} \\
   a^{12} &\equiv 1 \pmod{7}.
   \end{aligned}$$

   Thus, for an arbitrary integer a,

   $$\begin{aligned}
   a^{13} &\equiv a \pmod{2}, \\
   a^{13} &\equiv a \pmod{3}, \\
   a^{13} &\equiv a \pmod{5}, \\
   a^{13} &\equiv a \pmod{7}, \text{ and} \\
   a^{13} &\equiv a \pmod{13}.
   \end{aligned}$$

   It follows that $a^{13} \equiv a \pmod{2 \cdot 3 \cdot 5 \cdot 7 \cdot 13}$.

(c) For an odd integer a let $\gcd(a, 4080) = 1$, where $4080 = 15 \cdot 16 \cdot 17$. From Euler's Theorem, and the observation that

$$\phi(15) = 8, \ \phi(16) = 8, \text{ and } \phi(17) = 16,$$

it follows that

$$\begin{aligned}
a^8 &\equiv 1 \pmod{15}, \\
a^8 &\equiv 1 \pmod{16}, \text{ and} \\
a^8 &\equiv 1 \pmod{17}.
\end{aligned}$$

This yields

$$\begin{aligned}
a^{32} &\equiv 1 \pmod{15}, \\
a^{32} &\equiv 1 \pmod{16}, \text{ and} \\
a^{32} &\equiv 1 \pmod{17}.
\end{aligned}$$

Consequently,

$$\begin{aligned}
a^{33} &\equiv a \pmod{15}, \\
a^{33} &\equiv a \pmod{16}, \text{ and} \\
a^{33} &\equiv a \pmod{17}.
\end{aligned}$$

for any odd a, so that $a^{33} \equiv a \pmod{15 \cdot 16 \cdot 17}$.

3. Let $\gcd(a, 2^{15} - 2^3) = 1$, where $2^{15} - 2^3 = 5 \cdot 7 \cdot 8 \cdot 9 \cdot 13$. Since $\phi(5) = 4, \phi(7) = 6, \phi(8) = 4, \phi(9) = 6$ and $\phi(13) = 12$, Euler's Theorem yields

$$\begin{aligned}
a^4 &\equiv 1 \pmod{5}, \\
a^6 &\equiv 1 \pmod{7}, \\
a^4 &\equiv 1 \pmod{8}, \\
a^6 &\equiv 1 \pmod{9}, \text{ and} \\
a^{12} &\equiv 1 \pmod{13}
\end{aligned}$$

Hence,

$$\begin{aligned}
a^{12} &\equiv (a^4)^3 \equiv 1 \pmod{5}, \\
a^{12} &\equiv (a^6)^2 \equiv 1 \pmod{7}, \\
a^{12} &\equiv (a^4)^3 \equiv 1 \pmod{8}, \text{ and} \\
a^{12} &\equiv (a^6)^2 \equiv 1 \pmod{9}.
\end{aligned}$$

These congruences lead to $a^{15} \equiv a^3$ modulo 5, 7, 8, 9 and 13 (which still holds if $\gcd(a, 2^{15} - 2^3) \neq 1$ since both sides of the appropriate congruence would reduce to zero) and so

$$a^{15} \equiv a^3 \pmod{5 \cdot 7 \cdot 8 \cdot 9 \cdot 13}$$

for any a.

5. Suppose that $\gcd(n, m) = 1$ for positive integers n and m. Then $m^{\phi(n)} \equiv 1 \pmod{n}$ and $n^{\phi(m)} \equiv 1 \pmod{m}$. Hence

$$
\begin{aligned}
m^{\phi(n)} + n^{\phi(m)} &\equiv 1 + 0 \equiv 1 \pmod{n} \\
m^{\phi(n)} + n^{\phi(m)} &\equiv 0 + 1 \equiv 1 \pmod{m}
\end{aligned}
$$

and so, since $\gcd(n, m) = 1$, $m^{\phi(n)} + n^{\phi(m)} \equiv 1 \pmod{mn}$.

7. Since $\gcd(3, 10) = 1$ and $???\phi(10) = 4$, Euler's Theorem implies that $3^4 \equiv 1 \pmod{10}$. Hence,

$$3^{100} \equiv (3^4)^{25} \equiv 1^{25} \equiv 1 \pmod{10},$$

so that the units digit of 3^{100} is 1.

9. Since $\phi(77) = 60$, and working modulo 77,

$$
\begin{aligned}
2^{100000} &\equiv (2^{60})^{1666} \cdot 2^{40} \equiv 1^{16666} \cdot 2^{40} \\
&\equiv (2^{10})^4 \equiv (23)^4 \equiv 10^2 \equiv 23.
\end{aligned}
$$

11. For any prime p, since $\gcd(p, (p-1)!) = 1$,

     (a) $\tau(p!) = \tau(p)\tau((p-1)!) = 2\tau((p-1)!)$;

     (b) $\sigma(p!) = \sigma(p)\sigma((p-1)!) = (p+1)\sigma((p-1)!)$;

     (c) $\phi(p!) = \phi(p)\phi((p-1)!) = (p-1)\phi((p-1)!)$.

13. Let p be an odd prime. The $\phi(p) = p - 1$ integers

$$\pm 1, \pm 2, \pm 3, \ldots, \pm(p-1)/2$$

form a reduced set of residues modulo p, since

$$-1 \equiv p - 1, -2 \equiv p - 2, -3 \equiv p - 3, \ldots, -\frac{p-1}{2} \equiv \frac{p-1}{2} + 1.$$

7.4 Some Properties of the Phi-Function

1. For $n > 1$, let $n = 2^k N$, where $k \geq 0$ and N is odd. If n is odd (so that $k = 0$) and if $d|n$, then n/d is odd, whence $(-1)^{n/d} = -1$. Thus,

$$\sum_{d|n}(-1)^{n/d}\phi(n) = \sum_{d|n}\phi(n) = -n$$

by Gauss's Theorem. If n is even, then $k > 1$ and

$$
\begin{aligned}
\sum_{d|n}(-1)^{n/d}\phi(n) &= \sum_{d|2^{k-1}N}\phi(d) - \sum_{k|n}\phi(2^k d) \\
&= 2^{k-1}N - \phi(2^k)\sum_{d|n}\phi(d) \\
&= 2^{k-1}N - 2^{k-1}N = 0.
\end{aligned}
$$

3. Since the functions $F(n) = \sum_{d|n}\mu^2(d)/\phi(d)$ and $G(n) = n/\phi(n)$ are both multiplicative, to show that $F = G$ it is enough to show that they agree on a power of a prime. Thus, for a prime p,

$$
\begin{aligned}
F(p^k) &= \mu^2(1)/\phi(1) + \mu^2(p)/\phi(p) + \cdots + \mu^2(p^k)/\phi(p^k) \\
&= \mu^2(1)/\phi(1) + \mu^2(p)/\phi(p) \\
&= 1 + \frac{1}{p-1} = \frac{1}{1-\frac{1}{p}} = \frac{p^k}{\phi(p^k)} = G(p^k).
\end{aligned}
$$

5. For $n > 1$, let $n = p_1^{k_1}p_2^{k_2}\cdots p_r^{k_r}$. Then, using Problem 3 of Section 6.2:

(a)

$$
\begin{aligned}
\sum_{d|n}\mu(d)\phi(d) &= (1-\phi(p_1))(1-\phi(p_2))\cdots(1-\phi(p_r)) \\
&= (1-(p_1-1))(1-(p_2-1))\cdots(1-(p_r-1)) \\
&= (2-p_1)(2-p_2)\cdots(2-p_r).
\end{aligned}
$$

(b) The function $F(n) = \sum_{d|n}d\phi(d)$ is multiplicative and, for a prime p,

$$
\begin{aligned}
F(p^k) &= 1\phi(1) + p\phi(p) + p^2\phi(p^2) + \cdots + p^k\phi(p^k) \\
&= 1 + p(p-1) + p^2(p^2-p) + \ldots + p^k(p^k - p^{k-1})
\end{aligned}
$$

$$
\begin{aligned}
&= 1 - p + p^2 - p^3 + p^4 - \cdots - p^{2k-1} + p^{2k} \\
&= \frac{(-p)^{2k+1} - 1}{-p - 1} \\
&= \frac{p^{2k+1} + 1}{p + 1}.
\end{aligned}
$$

Thus

$$
\begin{aligned}
F(n) &= F(p_1^{k_1})F(p_2^{k_2}) \cdots F(p_r^{k_r}) \\
&= \left(\frac{p_1^{2k_1+1} + 1}{p_1 + 1} \right) \cdots \left(\frac{p_r^{2k_r+1} + 1}{p_r + 1} \right).
\end{aligned}
$$

(c) The function $G(n) = \sum_{d|n} \phi(d)/d$ is multiplicative, hence for a prime p,

$$
\begin{aligned}
G(p^k) &= \frac{\phi(1)}{1} + \frac{\phi(p)}{p} + \frac{\phi(p^2)}{p^2} + \cdots + \frac{\phi(p^k)}{p^k} \\
&= 1 + \frac{p-1}{p} + \frac{p^2 - p}{p^2} + \cdots + \frac{p^k - p^{k-1}}{p^k} \\
&= 1 + \frac{p-1}{p} + \frac{p-1}{p} + \cdots + \frac{p-1}{p} \\
&= 1 + k\frac{p-1}{p}.
\end{aligned}
$$

Thus,

$$
\begin{aligned}
G(n) &= G(p_1^{k_1})G(p_2^{k_2}) \cdots G(p_r^{k_r}) \\
&= \left(1 + \frac{k_1(p_1 - 1)}{p_1} \right) \cdots \left(1 + \frac{k_r(p_r - 1)}{p_r} \right).
\end{aligned}
$$

7. Let n be square-free, say $n = p_1 p_2 \ldots p_r$ for distinct primes p_i. Induct on r, the number of primes in n. For $r = 1$,

$$
\begin{aligned}
\sum_{d|p_1} \phi(dk^{k-1})\phi(d) &= \sigma(1)\phi(1) + \sigma(p_1^{k-1})\phi(p_1) \\
&= 1 + \left(\frac{p_1^k - 1}{p_1 - 1} \right)(p_1 - 1) = p_1^k.
\end{aligned}
$$

Assume the result holds for $r = j - 1$ and consider the case $r = j$, that is, $n = p_1 \ldots p_{j-1}p_j$. Then, for $n = Np_j$ where $N = p_1 \ldots p_{j-1}$,

$$
\sum_{d|n} \sigma(d^{k-1})\phi(d) = \sum_{d|N} \sigma(d^{k-1})\phi(d) + \sum_{d|N} \sigma((p_j d)^{k-1})\phi(p_j d)
$$

$$
\begin{aligned}
&= \sum_{d|N} \sigma(d^{k-1})\phi(d) + \sigma(p_j^{k-1})\phi(p_j)\sum_{d|N}\sigma(d^{k-1})\phi(d) \\
&= (1 + \sigma(p_j^{k-1})\phi(p_j))\sum_{d|N}\sigma(d^{k-1})\phi(d) \\
&= p_j^k \cdot N^k = (Np_j)^k = n^k.
\end{aligned}
$$

9. Since $3n + 2 \equiv -1 \pmod 3$, the prime factorization of $3n + 2$ must contain a prime $p \equiv -1 \pmod 3$ occurring to an odd power k. Thus

$$
\begin{aligned}
\sigma(p^k) &\equiv 1 + p + p^2 + \cdots + p^k \\
&\equiv 1 + (-1) + 1 + \cdots + (-1) \equiv 0 \pmod 3
\end{aligned}
$$

or $3|\sigma(p^k)$. From the multiplicative nature of σ, it follows that $3|\sigma(3n+2)$.

In the case of $4n + 3 \equiv -1 \pmod 4$, its prime factorization must involve a prime $q \equiv -1 \pmod 4$ to an odd power j. Then

$$
\begin{aligned}
\sigma(q^j) &\equiv 1 + q + q^2 + \cdots + q^j \\
&\equiv 1 + (-1) + 1 + \cdots + (-1) \equiv 0 \pmod q,
\end{aligned}
$$

whence $4|\sigma(q^j)$ and, in turn, $4|\sigma(4n+3)$.

11. If $d|n$, let $S_d = \{k\,|\,\gcd(k,n) = d; 1 \le k \le n\}$. The proof of Gauss's Theorem indicates that there are $\phi(n/d)$ integers in S_d and that the sets S_d partition $1, 2, ..., n$. Thus, $\gcd(k, n)$ takes on the value d for $\phi(n/d)$ choices of k, with $1 \le k \le n$; and so

$$
\sum_{k=1}^{n} \gcd(k, n) = \sum_{d|n} d\phi(n/d) = n\sum_{d|n}\phi(d)/d,
$$

where the last equality arises on replacing n/d by d.

13. Given $n \le 1$, let $k = n^2$. By Problem 10 of Section 7.2,

$$
\phi(k) = n\phi(n)
$$

whence $n|\phi(k)$.

15. Since the functions $F(n) = \sum_{d|n}\sigma(d)\phi(n/d)$ and $G(n) = n\tau(n)$ are both multiplicative, it is enough to show that they agree at p^k, a

power of a prime. Now

$$
\begin{aligned}
F(p^k) &= \sigma(1)\phi(p^k) + \sigma(p)\phi(p^{k-1}) + \sigma(p^2)\phi(p^{k-2}) + \cdots + \sigma(p^k)\phi(1) \\
&= (p^k - p^{k-1}) + \left(\frac{p^2 - 1}{p - 1}p^{k-2}(p - 1)\right) + \left(\frac{p^3 - 1}{p - 1}p^{k-3}(p - 1)\right) \\
&\quad + \cdots + \left(\frac{p^{k+1} - 1}{p - 1}\right) \\
&= (p^k - p^{k-1}) + (p^k - p^{k-2}) + (p^k - p^{k-3}) \\
&\quad + \cdots + (p^k + p^{k-1} + \cdots + p + 1) \\
&= (k + 1)p^k = G(p^k).
\end{aligned}
$$

To confirm that F actually is a multiplicative function, just mimic the argument of Theorem 6-8.

Similarly, it suffices to show that the functions $H(n) = \sum_{d|n} \tau(d)\phi(n/d)$ and $\sigma(n)$ agree when $n = p^k$, a power of a prime. Now

$$
\begin{aligned}
H(p^k) &= \tau(1)\phi(p^k) + \tau(p)\phi(p^{k-1}) + \tau(p^2)\phi(p^{k-2}) + \cdots + \tau(p^k)\phi(1) \\
&= (p^k - p^{k-1}) + 2(p^{k-1} - p^{k-2}) + 3(p^{k-2} - p^{k-3}) \\
&\quad + \cdots + k(p - 1) + (k + 1) \\
&= p^k + p^{k-1} + \cdots + p + 1 = \sigma(p^k).
\end{aligned}
$$

Chapter 8

Primitive Roots and Indices

8.1 The Order of an Integer Modulo n

1. (a) By Theorem 8-1, it suffices to consider exponents that divide 16. Modulo 17,

$$2^2 \equiv 4, \quad 2^4 \equiv 16, \quad 2^8 \equiv 1,$$
$$3^2 \equiv 9, \quad 3^4 \equiv 13, \quad 3^8 \equiv 16, \quad 3^{16} \equiv 1,$$
$$5^2 \equiv 8, \quad 5^4 \equiv 13, \quad 5^8 \equiv 16, \quad 5^{16} \equiv 1.$$

Thus 2, 3, 5 have orders 8, 16 and 16, respectively.

(b) Consider the divisors 2, 3, 6 and 9 of 18. Modulo 19,

$$2^2 \equiv 4, \quad 2^3 \equiv 8, \quad 2^6 \equiv 7, \quad 2^9 \equiv 18, \quad 2^{18} \equiv 1,$$
$$3^2 \equiv 9, \quad 3^3 \equiv 8, \quad 3^6 \equiv 7, \quad 3^9 \equiv 18, \quad 3^{18} \equiv 1,$$
$$5^2 \equiv 6, \quad 5^3 \equiv 11, \quad 5^6 \equiv 7, \quad 5^9 \equiv 1.$$

Hence 2,3 and 5 have orders 18, 18 and 9, respectively.

(c) Using the exponents 2, 11 and 22, and working modulo 23,

$$2^2 \equiv 4, \quad 2^{11} \equiv 1,$$
$$3^2 \equiv 9, \quad 3^{11} \equiv 1,$$
$$5^2 \equiv 2, \quad 5^{11} \equiv 22, \quad 5^{22} \equiv 1.$$

Thus 2,3 and 5 have orders 11,.11 and 22, respectively.

3. Clearly, $2^n \equiv 1 \pmod{2^n - 1}$. If $1 \le k < n$, then $2^k - 1 < 2^n - 1$, implying that $2^k \equiv 1 \pmod{2^n - 1}$; for, in the contrary case, $(2^n - 1)|(2^k - 1)$, which is impossible. Thus the order of 2 modulo $2^n - 1$ must be n. Then by Theorem 8-1, $n|\phi(2^n - 1)$.

5. Let a have order 3 modulo p, where p is a prime. Then $a^3 \equiv 1$ (mod p), so that $p|(a^3 - 1)$ or $p|(a - 1)(a^2 + a + 1)$. It follows that either $p|(a-1)$ or $p|(a^2 + a + 1)$. Since a has order 3 it cannot happen that $p|(a - 1)$; thus $p|(a^2 + a + 1)$, or $a^2 + a + 1 \equiv 0$ (mod p). This implies that $a^2 + 2a + 1 \equiv a$ (mod p), whence

$$(a + 1)^3 = (a + 1)^2(a + 1) \equiv a(a + 1) \equiv a^2 + a \equiv -1 \pmod{p},$$

and consequently $(a + 1)^6 \equiv 1$ (mod p). Since the order of $a + 1$ modulo p must divide 6, it must equal 1,2,3 or 6. Each of the first three cases leads to a contradiction:

if $a + 1 \equiv 1$ (mod p), then $p|a$;

if $(a + 1)^2 \equiv 1$ (mod p), then $a \equiv 1$ (mod p);

if $(a + 1)^3 \equiv 1$ (mod p), then $-1 \equiv$ (mod p) or $p = 2$.

Thus the order of $a + 1$ modulo p is 6.

7. Assume that there are only a finite number of primes of the form $4k + 1$; call them $p_1, p_2, \ldots, p_r$. By Problem 6(a), the odd integer $(2p_1 p_2 \cdots p_r)^2 + 1$ has a prime divisor q of the form $4k + 1$. Then $q = p_i$ for some $1 \le i \le r$. This leads to $q|1$, which is impossible. Similar arguments hold for primes of the form $6k + 1$ and $8k + 1$, upon considering integers of the type $(3p_1 p_2 \cdots p_r)^2 + (3p_1 p_2 \cdots p_r) + 1$ and $(2p_1 p_2 \cdots p_r)^4 + 1$.

9. Let p be an odd prime. Assume that there are only a finite number of primes of the form $2kp + 1$, say, $q_1, q_2, \ldots, q_r$. Put $a = 2q_1 q_2 \cdots q_r$ and consider the odd integer $a^{p-1} + a^{p-2} + \cdots + a + 1$. If q is an odd prime divisor of this integer, then $a^p \equiv 1$ (mod q). By Problem 8(a), either $q|a - 1$ or q is of the form $2kp + 1$. In the first case, $a \equiv 1$ (mod q), so that $a^{p-1} + a^{p-2} + \cdots + a + 1 \equiv p$ (mod q); since $q|p$ with p prime, it follows that $p = q$. But then $1 \equiv a \equiv 2(2k_1 p + 1) \cdots (2k_r p + 1) \equiv 2$ (mod p), a contradiction. Thus q is of the type $2kp + 1$ and hence q must be one of the q_i. This means that $q|a^j$ for $1 \le j \le p - 1$, and therefore $q|1$, which is impossible.

11. Let r be a primitive root of n. By Theorem 8-3, the order of r^k modulo n is $\phi(n)/\gcd(k, \phi(n))$. It follows that r^k is also a primitive root (that is, r^k has order $\phi(n)$) if and only if $\gcd(k, \phi(n)) = 1$.

13. (a) Assume that p and $q > 3$ are odd primes with $q|R_p$. Since $9R_p = 10^p - 1$, it follows that $q|10^p - 1$. By Problem 8(a), either $q|(10-1)$ or $q = 2kp + 1$ for some k; but $q > 3$, so the first possibility cannot hold.

(b) For $R_5 = 11111$, the prime divisors are of the form $10k + 1$ by part (a). Now 11 and 31 fail to divide R_5, but $R_5 = 41 \cdot 271$; hence, 41 is the smallest prime divisor of R_5. For $R_7 = 1111111$, the prime divisors are of the form $14k + 1$. The first prime of this type to divide R_7 is 239; that is, $R_7 = 239 \cdot 4649$.

8.2 Primitive Roots for Primes

1. (a) Let p be an odd prime and $1 \leq a \leq p - 1$ be a solution of $x^2 \equiv 1$ (mod p). Then $p|(a - 1)(a + 1)$, so that either $p|a - 1$ or $p|a + 1$. Then $a \equiv 1$ (mod p), or $a \equiv -1 \equiv p - 1$ (mod p); hence $a = 1$ or $a = p - 1$.

(b) Let p be an odd prime and $1 \leq a \leq p - 1$. By Fermat's Theorem, $a^{p-1} \equiv 1$ (mod p) or $p|(a - 1)(a^{p-2} + \cdots + a + 1)$; then either $p|a - 1$, in which case $a \equiv 1$ (mod p), or $p|a^{p-2} + \cdots + a + 1$, in which case, $a^{p-2} + \cdots + a + 1 \equiv 0$ (mod p). In the first instance $a = 1$: the remaining integers, namely $2, 3, \ldots, p - 1$ are the only solutions of $x^{p-2} + \cdots + x + 1 \equiv 0$ (mod p) by Lagrange's Theorem.

3. The integer 2 is a primitive root of 11. By the corollary to Theorem 8-4, the $\phi(10) = 4$ primitive roots of 11 are, modulo 11,

$$2^1 \equiv 2, \ 2^3 \equiv 8, \ 2^7 \equiv 7, 2^9 \equiv 6.$$

Since 2 is a primitive root of 19, the $\phi(18) = 6$ primitive roots of 19 are, modulo 19,

$$2^1 \equiv 2, 2^5 \equiv 13, 2^7 \equiv 14, 2^{11} \equiv 15, 2^{13} \equiv 3, 2^{17} \equiv 10.$$

The integer 5 is a primitive root of 23, and there are $\phi(22) = 10$ in all. Modulo 23, these are

$$5^1 \equiv 5, \quad 5^3 \equiv 10, \quad 5^5 \equiv 20, \quad 5^7 \equiv 17, \quad 5^9 \equiv 11,$$
$$5^{13} \equiv 21, \quad 5^{15} \equiv 19, \quad 5^{17} \equiv 15, \quad 5^{19} \equiv 7, \quad 5^{21} \equiv 14.$$

5. Since 2 is a primitive root of 61, $2k$ will have order 4 modulo 61 when $60/\gcd(k, 60)$ equals 4; that is, when $\gcd(k, 60) = 15$. This occurs when $k = 15$ and when $k = 45$. The integers having order 4 are, modulo 61,

$$2^{15} \equiv 11 \quad \text{and} \quad 2^{45} \equiv 50.$$

7. Let r be a primitive root of the prime $p > 3$, and put $r' = r^{p-2}$. Then $rr' = r \cdot r^{p-2} = r^{p-1} \equiv 1 \pmod{p}$. Hence, by Problem 6(c), r' is also a primitive root of p; $r \not\equiv r' \pmod{p}$, for otherwise $p = 3$.

9. Assume that r is a primitive root of the prime p. By Problem 6(a), $r^{(p-1)/2} \equiv -1 \pmod{p}$ or

$$[r^{(p-1)/4}]^2 + 1 \equiv 0 \pmod{p}.$$

Thus, $r^{(p-1)/4}$ satisfies the congruence $x^2 + 1 \equiv 0 \pmod{p}$.

11. Let r be a primitive root of the prime p. There are $\phi(p-1)$ primitive roots of p and these have the form r^k, where $\gcd(k, p-1) = 1$ and $1 \le k \le p-1$. Denote these $\phi(p-1)$ values of k by $k_1, k_2, \ldots, k_{\phi(p-1)}$. Then the product of the primitive roots is congruent modulo p to r^k, where

$$k = k_1 + k_2 + \cdots + k_{\phi(p-1)} = \frac{(p-1)\phi(p-1)}{2}$$

by Theorem 7-7. It follows that the product of the primitive roots is given by

$$r^{(p-1)\phi(p-1)/2} = [r^{(p-1)/2}]^{\phi(p-1)} \equiv (-1)^{\phi(p-1)} \pmod{p}.$$

8.3   Composite Numbers Having Primitive Roots

1. (a) There are $\phi(\phi(26)) = \phi(12) = 4$ primitive roots of 26. Since 2, 6, 7 and 11 are primitive roots of 13, the integers $2 + 13 = 15$, $6 + 13 = 19$, 7 and 11 are the primitive roots of 26. There are $\phi(\phi(15)) = \phi(20) = 8$ primitive roots of 25. The other primitive roots of 25 are of the form $2k$, where $\gcd(k, \phi(25)) = 1$. These are, modulo 25,

$$2, \qquad 2^3 \equiv 8, \quad 2^7 \equiv 3, \quad 2^9 \equiv 12,$$
$$2^{11} \equiv 23, \quad 2^{13} \equiv 17, \quad 2^{17} \equiv 23, \quad 2^{19} \equiv 13.$$

(b) The integer 2 is a primitive root of 3^2, 3^3 and 3^4. The $\phi(\phi(32)) = \phi(6) = 2$ primitive roots of 3^2 are 2 and $2^5 \equiv 5 \pmod 9$. The $\phi(\phi(3^3)) = \phi(18) = 6$ primitive roots of 3^3 are, modulo 27,

$$2, \qquad 2^5 \equiv 5, \qquad 2^7 \equiv 20,$$
$$2^{11} \equiv 23, \quad 2^{13} \equiv 11, \quad 2^{17} \equiv 14.$$

The $\phi(\phi(3^4)) = \phi(54) = 18$ primitive roots of 34 are, modulo 81,

$$2, \qquad 2^5 \equiv 32, \quad 2^7 \equiv 47, \quad 2^{11} \equiv 23, \quad 2^{13} \equiv 11, \quad 2^{17} \equiv 14,$$
$$2^{19} \equiv 56, \quad 2^{23} \equiv 5, \quad 2^{25} \equiv 20, \quad 2^{29} \equiv 77, \quad 2^{31} \equiv 65, \quad 2^{35} \equiv 68,$$
$$2^{37} \equiv 29, \quad 2^{41} \equiv 59, \quad 2^{43} \equiv 74, \quad 2^{47} \equiv 50, \quad 2^{49} \equiv 38, \quad 2^{53} \equiv 41.$$

3. Assume that r is a primitive root of p^2, where p is an odd prime. Each of the integers r^p, r^{2p}, $\ldots, r^{(p-1)p}$ satisfies the congruence $x^{p-1} \equiv 1 \pmod{p^2}$, since

$$\left(r^{kp}\right) = r^{k\phi(p^2)} = \left(r^{\phi(p^2)}\right)^k \equiv 1^k \equiv 1 \pmod{p^2}$$

for $1 \le k \le p - 1$. By Theorem 8-5, the congruence has at most $p - 1$ solutions, hence r^p, $r^{2p}, \ldots, r^{(p-1)p}$ provide all the solutions. Notice that if $r^{kp} \equiv r^{jp} \pmod{p^2}$, with $1 \le j < k \le p - 1$, then $r^{(k-j)p} \equiv 1 \pmod{p^2}$; hence, $\phi(p^2)|(k-j)p$, or $(p-1)|(k-j)$ implying that $k = j$.

5. There are $\phi(\phi(41)) = \phi(40) = 16$ primitive roots of 41. Since 6 is a primitive root of 41, the other primitive roots are of the form 6^k, where $\gcd(k, 40) = 1$. Modulo 41, they are

$$6, \qquad 6^3 \equiv 11, \quad 6^7 \equiv 29, \quad 6^9 \equiv 19,$$
$$6^{11} \equiv 28, \quad 6^{13} \equiv 24, \quad 6^{17} \equiv 26, \quad 6^{19} \equiv 34,$$
$$6^{21} \equiv 35, \quad 6^{23} \equiv 30, \quad 6^{27} \equiv 12, \quad 6^{29} \equiv 22,$$
$$6^{31} \equiv 13, \quad 6^{33} \equiv 17, \quad 6^{37} \equiv 15, \quad 6^{39} \equiv 7.$$

The $\phi(\phi(82)) = \phi(\phi(2)\phi(41)) = 16$ primitive roots of 82 will be the odd primitive roots of 41, together with the even primitive roots of 41 with 41 added to each of them. They are

7, 11, 13, 15, 17, 19, 29, 35,
$6 + 41 = 47$, $12 + 41 = 53$, $22 + 41 = 63$, $24 + 41 = 65$,
$26 + 41 = 67$, $28 + 41 = 69$, $30 + 41 - 71$, $34 + 41 = 75$.

7. Assume that r is a primitive root of the odd prime p, and consider $r + tp$. Since $r + tp \equiv r \pmod{p}$, $r + tp$ is also a primitive root of p. Let n be the order of $r + tp$ modulo p^k, where $k > 1$. Then $n | \phi(p^k)$ or $n | p^{k-1}(p-1)$. If $n \neq p^{k-1}(p-1)$, then $n | \phi^{k-2}(p-1)$, leading to

$$(r + tp)^{p^{k-2}(p-1)} \equiv 1 \pmod{p^k}.$$

But this contradicts the lemma preceding Theorem 8-9. Thus $n = p^{k-1}(p-1)$, which makes $r + tp$ a primitive root of p^k for $k > 1$.

9. If $n = 5040 = 2^4 \cdot 3^2 \cdot 5 \cdot 7$, then

$$\begin{aligned} \lambda(n) &= \operatorname{lcm}(\lambda(2^4), \phi(3^2), \phi(5), \phi(7)) \\ &= \operatorname{lcm}(4, 6, 4, 6) = 12. \end{aligned}$$

But $\phi(n) = \phi(2^4)\phi(3^2)\phi(5)\phi(7) = 8 \cdot 6 \cdot 4 \cdot 6 = 1152$.

11. (a) Let $\gcd(a, n) = 1$. To see that $x \equiv ba^{\lambda(n)-1} \pmod{n}$ is a solution of the linear congruence $ax \equiv b \pmod{n}$, notice that

$$a(ba^{\lambda(n)-1}) = a^{\lambda(n)}b \equiv 1 \cdot b \equiv b \pmod{n}$$

by Problem 8(c).

(b) Consider the linear congruence $13x \equiv 2 \pmod{40}$. Here, $\lambda(40) = \operatorname{lcm}(\lambda(2^3), \phi(5)) = \operatorname{lcm}(2, 4) = 4$, so that

$$x \equiv 2 \cdot 13^3 = 34 \pmod{40}$$

is a solution. For the congruence $3x \equiv 13 \pmod{77}$, $\lambda(77) = \operatorname{lcm}(\phi(7), \phi(11)) = \operatorname{lcm}(6, 10) = 30$. Hence, a solution of $3x \equiv 13 \pmod{77}$ is given by

$$x \equiv 13 \cdot 3^{29} \equiv 30 \pmod{77}.$$

8.4   The Theory of Indices

1. Since 2 is a primitive root of 13, the $\phi(12) = 4$ primitive roots of 13 are, modulo 13,

$$2^1 \equiv 2, 2^5 \equiv 6, 2^7 \equiv 11, 2^{11} \equiv 7.$$

The congruences

$$\begin{aligned} 2^9 &\equiv 5 \pmod{13}, & 6^9 &\equiv 5 \pmod{13}, \\ 7^3 &\equiv 5 \pmod{13}, & 11^3 &\equiv 5 \pmod{13} \end{aligned}$$

imply that $\operatorname{ind}_2 5 = 9$, $\operatorname{ind}_6 5 = 9$, $\operatorname{ind}_7 5 = 3$ and $\operatorname{ind}_{11} 5 = 3$.

3. (a) Consider the congruence $x^{12} \equiv 13 \pmod{17}$. Taking indices with respect to the primitive root 3 yields $12\,\mathrm{ind}\,x \equiv \mathrm{ind}\,13 \pmod{16}$, or $12\,\mathrm{ind}\,x \equiv 4 \pmod{16}$. This last congruence has solutions $\mathrm{ind}\,x = 3, 7, 11$ and $15 \pmod{16}$, which leads to $x \equiv 10, 11, 7$ and $6 \pmod{17}$.

   (b) For the congruence $8x^5 \equiv 10 \pmod{17}$, we have $\mathrm{ind}\,8 + 5\,\mathrm{ind}\,x \equiv \mathrm{ind}\,10 \pmod{16}$, or $10 + 5\,\mathrm{ind}\,x \equiv 3 \pmod{16}$. Thus $5\,\mathrm{ind}\,x \equiv 9 \pmod{16}$, with solution $\mathrm{ind}\,x \equiv 5 \pmod{16}$. Then $x \equiv 5 \pmod{17}$.

   (c) If $9x^8 \equiv 8 \pmod{17}$, then $\mathrm{ind}\,9 + 8\,\mathrm{ind}\,x \equiv \mathrm{ind}\,8 \pmod{16}$. It follows that $2 + 8\,\mathrm{ind} \equiv 10 \pmod{16}$, or $8\,\mathrm{ind}\,x \equiv 8 \pmod{16}$. This congruence has solutions $x \equiv 1, 3, 5, 7, 9, 11$ and $15 \pmod{16}$ so that $x \equiv 3, 10, 5, 11, 14, 7, 12$ and $6 \pmod{17}$.

   (d) Let $7^x \equiv 7 \pmod{17}$. Then $x\,\mathrm{ind}\,7 \equiv \mathrm{ind}\,7 \pmod{16}$ or $x \equiv 1 \pmod{16}$.

7. Let r be a primitive root of the odd prime p. By Problem 6(a) of Section 8.2, $r^{(p-1)/2} \equiv -1 \pmod{p}$. If $r^k \equiv -1 \pmod{p}$, where $0 < k < (p-1)/2$, then $r^{2k} \equiv 1 \pmod{p}$ with $2k < p - 1$; hence r would not be a primitive root of p. This implies that $\mathrm{ind}_r(-1) = (p-1)/2$. Since $p - 1 \equiv -1 \pmod{p}$, $\mathrm{ind}_r(p-1) = (p-1)/2$ also.

9. (a) For an odd prime p, the corollary to Theorem 8-12 implies that $x^2 \equiv -1 \pmod{p}$ has a solution if and only if $(-1)^{(p-1)/2} \equiv 1 \pmod{p}$; that is, if and only if $p \equiv 1 \pmod{4}$.

   (b) For an odd prime p, the congruence $x^4 \equiv -1 \pmod{p}$ has a solution if and only if $(-1)^{(p-1)/d} \equiv 1 \pmod{p}$, where $d = \gcd(4, p-1)$. This condition holds if and only if $p \equiv 1 \pmod{8}$, since $(p-1)/d$ is odd when $p \equiv 3,\ 5$ or $7 \pmod{8}$.

11. Since $\gcd(3, 18) = 3$ and $3^6 \not\equiv 1 \pmod{19}$, the congruence $x^3 \equiv 3 \pmod{19}$ has no solution by Theorem 8-12. But $11^6 \equiv 1 \pmod{19}$ so that $x^3 \equiv 11 \pmod{19}$ has three solutions.

13. Assume that p is a prime and $\gcd(k, p-1) = 1$. For each value of a, where $1 \le a \le p-1$, Theorem 8-12 implies that the congruence $x^k \equiv a \pmod{p}$ has a unique solution b which is not congruent to zero modulo p. Since $b^k \equiv a \pmod{p}$, we have $\gcd(b^k, p) = \gcd(a, p) = 1$ by Problem 3 of Section 4.2. As a runs through the values $1, 2, \ldots, p-1$, so does b; thus the integers $1^k, 2^k, \ldots, (p-1)^k$ form a reduced set of residues modulo p.

15. Suppose that r is a primitive root of the odd prime p. Taking indices relative to r,

$$\begin{aligned} r^{\operatorname{ind}(p-1)} &\equiv p - a \equiv (-1)a \equiv r^{(p-1)/2}r^{\operatorname{ind} a} \\ &\equiv r^{(p-1)/2+\operatorname{ind} a} \pmod{p}. \end{aligned}$$

From Theorem 8-2, this yields

$$\operatorname{ind}(p-a) \equiv \operatorname{ind} a + (p-1)/2 \pmod{p-1}.$$

17. Given the congruence $9^x \equiv b \pmod{13}$, take indices with respect to the primitive root 2 to obtain $x(\operatorname{ind} 9) \equiv \operatorname{ind} b \pmod{12}$, or $8x \equiv \operatorname{ind} b \pmod{12}$. There will be a solution, provided $\gcd(8,12)|\operatorname{ind} b$; that is, whenever $4|\operatorname{ind} b$. Hence, $\operatorname{ind} b = 4$, 8 or 12, which implies that $b \equiv 3$, 9 or 1 $\pmod{13}$.

Chapter 9

The Quadratic Reciprocity Law

9.1 Euler's Criterion

1. (a) The quadratic congruence $x^2 + 7x + 10 \equiv 0 \pmod{11}$ can be rewritten as $(2x + 7)^2 \equiv 9 \pmod{11}$; or, as $y^2 \equiv 9 \pmod{11}$, where $y \equiv 2x + 7 \pmod{11}$. Now $y \equiv 3 \pmod{11}$ and $y \equiv -3 \equiv 8 \pmod{11}$. The linear congruence $2x + 7 \equiv 3 \pmod{11}$ has solution $x \equiv 9 \pmod{11}$, while $2x + 7 \equiv 8 \pmod{11}$ has $x \equiv 6 \pmod{11}$ as its solution.

   (b) The congruence $3x^2 + 9x + 7 = 0 \pmod{13}$ implies that $(6x + 9)^2 \equiv 10 \pmod{13}$, or rather, $y^2 \equiv 10 \pmod{13}$ where $y \equiv 6x + 9 \pmod{13}$. Since $y \equiv \pm 6 \pmod{13}$, $6x + 9 \equiv 6 \pmod{13}$ gives $x \equiv 6 \pmod{13}$ and $6x + 9 \equiv -6 \pmod{13}$ yields $x \equiv 4 \pmod{13}$.

   (c) Rewrite the congruence $5x^2 + 6x + 1 \equiv 0 \pmod{23}$ as $(10x + 6)^2 \equiv 16 \pmod{23}$; that is, as $y^2 \equiv 16 \pmod{23}$), where $y \equiv 10x + 6 \pmod{23}$. Clearly $y = \pm 4 \pmod{23}$. When $y \equiv 4 \pmod{23}$, the congruence $10x + 6 \equiv 4 \pmod{23}$ implies that $x \equiv 9 \pmod{23}$; when $y \equiv -4 \pmod{23}$, the congruence $10x + 6 \equiv -4 \pmod{23}$ yields $x \equiv 22 \pmod{23}$.

3. (a) Let p be an odd prime, and let the integer n satisfy $1 \leq n \leq (p-1)/2$. Then n^2 is a quadratic residue of p by Euler's Criterion; for, using Euler's Theorem, $(n^2)^{(p-1)/2} \equiv n^{p-1} \equiv 1 \pmod{p}$. On the other hand, suppose that a is a quadratic residue of p. If r is

a primitive root of p, then the proof of Theorem 9-1 tells us that $a \equiv r^{2j} \equiv (r^j)^2 \pmod{p}$ for some j with $1 \leq j \leq (p-1)/2$. Now r^j is congruent modulo p to one of $\pm 1, \pm 2, \ldots, \pm(p-1)/2$, so that a is congruent to one of the integers $1^2, 2^2, \ldots, ((p-1)/2)^2$.

(b) From part (a), the quadratic residues of 17 are, modulo 17:

$$1^2 \equiv 1, \quad 2^2 \equiv 4, \quad 3^2 \equiv 9, \quad 4^2 \equiv 16,$$
$$5^2 \equiv 8, \quad 6^2 \equiv 2, \quad 7^2 \equiv 15, \quad 8^2 \equiv 13.$$

5. (a) If a is a quadratic residue of the odd prime p, then $a^{(p-1)/2} \equiv 1 \pmod{p}$ by Euler's Criterion. It follows that the order of a modulo p is less than $p-1 = \phi(p)$, hence a cannot be a primitive root of p.

(b) Let a be a quadratic residue of the odd prime p. Then

$$(p-a)^{(p-1)/2} \equiv (-a)^{(p-1)/2} \equiv (-1)^{(p-1)/2} a^{(p-1)/2}$$
$$\equiv (-1)^{(p-1)/2} \pmod{p}.$$

Thus $(p-a)^{(p-1)/2}$ is congruent modulo p to 1 or -1 according as $p \equiv 1 \pmod 4$ or $p \equiv 3 \pmod 4$; that is, $p-a$ is a quadratic residue, or nonresidue, of p according as $p \equiv 1 \pmod 4$ or $p = 3 \pmod 4$.

(c) Suppose that a is a quadratic residue of the prime $p \equiv 3 \pmod 4$. Then

$$(\pm a^{(p+1)/4})^2 \equiv a^{(p+1)/2} \equiv a^{(p-1)/2} \cdot a \equiv 1 \cdot a \pmod{p},$$

making $\pm a^{(p+1)/4}$ solutions of the quadratic congruence $x^2 \equiv a \pmod{p}$.

7. Let a be a quadratic nonresidue of the prime p, where $p = 2^k + 1$ for some $k \geq 1$. Then, by Euler's Criterion, $a^{2^{k-1}} = a^{(p-1)/2} \equiv -1 \pmod{p}$; hence upon squaring, $a^{2^k} \equiv 1 \pmod{p}$. This means that the order of a modulo p is a divisor of 2^k. If the order of a is 2^j, with $j < k$, then repeated squaring would lead to $a^{2^{k-1}} \equiv 1 \pmod{p}$, or $1 \equiv -1 \pmod{p}$, so that $p = 2$. This is impossible, so that a has order $2^k = p - 1 = \phi(p)$, making a a primitive root of p.

9. (a) Let $ab \equiv r \pmod{p}$, where r is a quadratic residue of the odd prime p. If one of a and b is a quadratic residue of p, say a, while

the other is a nonresidue, then

$$r^{(p-1)/2} \equiv (ab)^{(p-1)/2} \equiv a^{(p-1)/2}b^{(p-1)/2}$$
$$\equiv 1 \cdot (-1) \equiv -1 \pmod{p}$$

by Euler's Criterion. This means that r cannot be a quadratic residue of p. Thus a and b are either both residues, or both non-residues.

(b) Let a and b be either both quadratic residues or both nonresidues of the odd prime p; then $\gcd(a,p) = \gcd(b,p) = 1$. If a' is a solution of the linear congruence $ax \equiv 1 \pmod{p}$, then multiplying the congruence $ax^2 \equiv b \pmod{p}$ by a' leads to $x^2 \equiv a'b \pmod{p}$. Assuming a and b are both quadratic residues of p,

$$(a'b)^{(p-1)/2} \equiv (a')^{(p-1)/2}b^{(p-1)/2} \equiv 1 \cdot 1 \equiv 1 \pmod{p},$$

since a' is a quadratic residue of p by part (a). This implies that $a'b$ is a quadratic residue of p, whence $x^2 \equiv a'b \pmod{p}$ is solvable: and, in turn, $ax^2 \equiv b \pmod{p}$ has a solution. A similar argument holds when a and b are both nonresidues of p.

11. (a) Since 2 is a primitive root of 19, the proof of Theorem 9-1 indicates that 2^k is a quadratic residue of 19 if and only if k is even. Modulo 19, these quadratic residues are

$$2^2 \equiv 4, \quad 2^4 \equiv 16, \quad 2^6 \equiv 7, \quad 2^8 \equiv 9, \quad 2^{10} \equiv 17,$$
$$2^{12} \equiv 11, \quad 2^{14} \equiv 6, \quad 2^{16} \equiv 5, \quad 2^{18} \equiv 1.$$

(b) The integer 2 is a primitive root of 29, while 3 is a primitive root of 31. Modulo 29, the quadratic residues of 29 are

$$2^2 \equiv 2, \quad 2^4 \equiv 16, \quad 2^6 \equiv 6, \quad 2^8 \equiv 24, \quad 2^{10} \equiv 9,$$
$$2^{12} \equiv 7, \quad 2^{14} \equiv 28, \quad 2^{16} \equiv 25, \quad 2^{18} \equiv 13, \quad 2^{20} \equiv 23,$$
$$2^{22} \equiv 5, \quad 2^2 \equiv 20, \quad 2^{26} \equiv 22, \quad 2^{28} \equiv 1.$$

The quadratic residues of 31 are, modulo 31,

$$3^2 \equiv 9, \quad 3^4 \equiv 19, \quad 3^6 \equiv 16, \quad 3^8 \equiv 20, \quad 3^{10} \equiv 25,$$
$$3^{12} \equiv 8, \quad 3^{14} \equiv 10, \quad 3^{16} \equiv 28, \quad 3^{18} \equiv 4, \quad 3^{20} \equiv 5,$$
$$3^{22} \equiv 14, \quad 3^{24} \equiv 2, \quad 3^{26} \equiv 18, \quad 3^{28} \equiv 7, \quad 3^{30} \equiv 1.$$

13. For the integers $a = 7$ and $n = 15$, we have $\gcd(7,15) = 1$ and $\phi(15) = 8$. But the quadratic congruence $x^2 \equiv 7 \pmod{15}$ has no solution, even though $7^{\phi(15)/2} \equiv 7^4 \equiv 1 \pmod{15}$. Thus the conditions of Problem 12 are not sufficient for the existence of a quadratic residue of n.

9.2   The Legendre Symbol and Its Properties

1. (a) $(19/23) = (-4/23) = (4/23)(-1/23) = 1 \cdot (-1) = -1$.

   (b) $(-23/59) = (36/59) = 1$

   (c) $(20/31) = (4/31)(5/31) \equiv 1 \cdot 5^{15} \equiv 1 \pmod{31}$, hence $(20/32) = 1$.

   (d) $(18/43) = (9/43)(2/43) \equiv 1 \cdot 2^{21} \equiv -1 \pmod{43}$, so that $(18/43) = -1$.

   (e) $(-72/131) = (-1/131)(36/131)(2/131) \equiv (-1) \cdot 1 \cdot 2^{65} \equiv 1 \pmod{131}$ and so $(-72/131) = 1$.

3. If p is an odd prime, then there are $\phi(p-1)$ primitive roots of p and these are all quadratic nonresidues of p. By Theorem 9-4 there are $(p-1)/2$ nonresidues in all, so that the number of nonresidues which are not primitive roots is $(p-1)/2 - \phi(p-1)$.

5. Consider primes of the form $p = 3 \cdot 2^n + 1$. When $n = 1$, we have $p = 7$; since $(2/7) = 1$, 2 is not a primitive root of 7. When $n = 2$, $p = 13$ and 2 is a primitive root of 13. Assume $n \geq 3$. Then $p = 8 \cdot 3 \cdot 2^{n-3} + 1 \equiv 1 \pmod{8}$, hence $(2/p) = 1$ by Theorem 9-6. This means that 2 is a quadratic residue of p, and consequently not a primitive root of p.

7. Let p be an odd prime and $\gcd(a, p) = 1$. If a' satisfies the congruence $ax \equiv 1 \pmod{p}$ then

$$a(a + 1) \equiv a(a + aa') \equiv a^2(1 + a') \pmod{p}.$$

It follows that

$$(a(a + 1)/p) = (a^2/p)((1 + a')/p) = ((1 + a')/p).$$

As a runs through the values 1 to $p - 2$ so does a', and consequently $1 + a'$ runs through the values 2 to $p - 1$. Hence

$$\sum_{a=1}^{p-2}(a(a+1)/p) = \sum_{a'=2}^{p-1}((1+a')/p) = \sum_{a=2}^{p-1}(a/p)$$

$$= \sum_{a=1}^{p-1}(a/p) - 1 = 0 - 1 = -1$$

by Theorem 9-4.

9. Assume that $p \equiv 7 \pmod 8$ is prime. Then, by Theorem 9-6, $(2/p) = 1$ or $2^{(p-1)/2} \equiv 1 \pmod p$. This says that $p | 2^{(p-1)/2} - 1$.

11. (a) Suppose that p and $q = 4p + 1$ are both primes and that a is a quadratic nonresidue of q. Since $\phi(q) = 4p$, the order of a modulo q is either 1, 2, 4, p, $2p$, or $4p$. But, because a is a quadratic nonresidue of q, $a^{2p} \equiv -1 \pmod q$. If a had order 1, 2, or p, it would follow that $1 \equiv -1 \pmod q$ or $q = 2$, which is impossible. Thus the order of a modulo q is either 4 or $4p$, the latter case implying that a is a primitive root of q.

    (b) Let p and $q = 4p + 1$ be (odd) primes. Then $q \equiv 5 \pmod 8$, so that $(2/q) = -1$, making 2 a quadratic nonresidue of q. By part (a), either 2 has order 4 modulo q or 2 is a primitive root of q. If 2 had order 4, then $2^4 \equiv 1 \pmod q$ or $q | 15$, so that $q = 3$ or $q = 5$; but neither 3 nor 5 is a prime of the form $4p + 1$. Thus 2 is a primitive root of q.

13. Let r be a primitive root of the odd prime p, so that $r^{(p-1)/2} \equiv -1 \pmod p$. According to Problem 12, the product of the quadratic residues of p is congruent to

$$r^{(p^2-1)/4} = (r^{(p-1)/2})^{(p+1)/2} \equiv (-1)^{(p+1)/2} \pmod p,$$

where $(-1)^{(p+1)/2}$ is either 1 or -1 according as $p \equiv 1 \pmod 4$ or $p \equiv 3 \pmod 4$.

15. Suppose that $p > 5$ is prime. Then the three integers 2, 5 and $10 = 2 \cdot 5$ are incongruent modulo p; and, by Problem 6(a), at least one is a quadratic residue of p. If 2 is a quadratic residue of p, then 1 and 2 are consecutive residues. If 5 is a quadratic residue of p, then 4 and 5 are consecutive residues. Similarly, if 10 is a quadratic residue of p, then 9 and 10 are consecutive residues.

17. Assume that p is an odd prime divisor of $9^n + 1$, so that $9^n \equiv -1 \pmod p$. Then

$$(-1/p) = (9^n/p) = ((3^n)^2/p) = 1$$

From Theorem 9-1, it follows that $p \equiv 1 \pmod 4$.

9.3 Quadratic reciprocity

1. (a) $(71/73) = (73/71) = (2/71) = 1$

    (b)

$$
\begin{aligned}
(-219/383) &= (-1/383)(3/383)(73/383) = (-1) \cdot 1 \cdot (383/73) \\
&= (18/73) = -(9/73)(2/73) = (-1) \cdot 1 \cdot 1 = -1
\end{aligned}
$$

    (c)

$$
\begin{aligned}
(461/773) &= (773/461) = (312/461) \\
&= (4/461)(3/461)(2/461)(13/461) \\
&= 1 \cdot (-1) \cdot (-1)(13/461) = (461/13) = (6/13) \\
&= (2/13)(3/13) = -1 \cdot 1 = -1
\end{aligned}
$$

    (d)

$$
\begin{aligned}
(1234/4567) &= (2/4567)(617/4567) = 1 \cdot (4567/617) \\
&= (248/617) = (4/617)(2/617)(31/617) \\
&= 1 \cdot 1 \cdot (617/31) = (28/31) = (4/31)(7/31) \\
&= 1 \cdot (-1)(31/7) = -(3/7) = -(-1) = 1
\end{aligned}
$$

    (e)

$$
\begin{aligned}
(3658/12703) &= (2/12703)(31/12703)(59/12703) \\
&= 1 \cdot (-1)(12703/31) \cdot (-1)(12703/59) \\
&= (24/31)(18/59) = (4/31)(3/31)(2/31)(9/59)(2/59) \\
&= 1 \cdot (-1) \cdot 1 \cdot 1 \cdot (-1) = 1
\end{aligned}
$$

3. (a) The quadratic congruence $x^2 \equiv 219 \pmod{419}$ is solvable since 219 is a quadratic residue of 419:

$$
\begin{aligned}
(219/419) &= (3/419)(73/419) = 1 \cdot (419/73) = (54/73) \\
&= (9/73)(3/73)(2/73) = 1 \cdot 1 \cdot 1 = 1.
\end{aligned}
$$

    (b) Consider the quadratic congruence $3x^2 + 6x + 5 \equiv 0 \pmod{89}$ or equivalently, $(x + 1)^2 \equiv 29 \pmod{89}$. This congruence is not solvable, since 29 is a quadratic nonresidue of 89:

$$
(29/89) = (89/29) = (2/29) = -1.
$$

(c) The quadratic congruence $2x^2 + 5x - 9 \equiv 0 \pmod{101}$ is equivalent to $(4x + 5)^2 \equiv 97 \pmod{101}$. Now

$$(97/101) = (101/97) = (4/97) = 1$$

and so 97 is a quadratic residue of 101; hence, $(4x + 5)^2 \equiv 97 \pmod{101}$ has a solution.

5. (a) For a prime $p > 3$, Theorems 9-2 and 9-10 imply that

if $p \equiv 1 \pmod{12}$, then $(-1/p) = 1$ and $(3/p) = 1$;
if $p \equiv 5 \pmod{12}$, then $(-1/p) = 1$ and $(3/p) = -1$;
if $p \equiv 7 \pmod{12}$, then $(-1/p) = -1$ and $(3/p) = -1$;
if $p \equiv 11 \pmod{12}$, then $(-1/p) = -1$ and $(3/p) = 1$;

Thus $(-3/p) = (-1/p)(3/p)$ has the value 1 when $p \equiv 1$ or 7 $\pmod{12}$, which is to say $p \equiv 1 \pmod 6$; and $(-3/p)$ has the value -1 when $p \equiv 5$ or 11 $\pmod{12}$, or rather $p \equiv 5 \pmod 6$.

(b) Assume that there are only a finite number of primes of the form $6k+1$; call them $p_1, p_2, \ldots, p_r$. Now the integer $N = (2p_1 p_2 \cdots p_r)^2 + 3$ has an odd prime divisor $p > 3$. Since $(2p_1 p_2 \cdots p_r)^2 \equiv -3 \pmod p$, we have $(-3/p) = 1$. By part (a) this implies that $p \equiv 1 \pmod 6$, whence $p = p_i$ for some $1 \leq i \leq r$. But then $p | 3$, which is impossible.

7. Suppose that there are only a finite number of primes of the form $8k + 3$; call them $p_1, p_2, \ldots, p_r$ and consider the odd integer $N = (p_1 p_2 \cdots p_r)^2 + 2$. Now N has an odd prime divisor p, so that $(p_1 p_2 \cdots p_r)^2 \equiv -2. \pmod p$. This implies that $(-2/p) = 1$, whence $p \equiv 1$ or 3 $\pmod 8$ by Problem 4. If all the prime divisors of N are of the form $8k + 1$, then $N \equiv 1 \pmod 8$; but this is not the case. Thus, N has a prime divisor $p \equiv 3 \pmod 8$, so that $p = p_i$ for some $1 \leq i \leq r$. It follows that $p | 2$, which is impossible.

9. Let p and q be odd primes satisfying $p = q + 4a$ for some a. Then $p \equiv 4a \pmod q$ and $q \equiv -4a \pmod p$. Since $p \equiv q \pmod 4$, there are two cases to consider. If $p \equiv 1 \pmod 4$ and $q \equiv 1 \pmod 4$, then from the Quadratic Reciprocity Law,

$$\begin{aligned}
(a/p) &= (-1/p)(4/p)(a/p) = (-4a/p) = (q/p) = (p/q) \\
&= (4a/q) = (4/q)(a/q) = (a/q).
\end{aligned}$$

If $p \equiv 3$ (mod 4) and $q \equiv 3$ (mod 4), then

$$
\begin{aligned}
(a/p) &= -(-1/p)(4/p)(a/p) = -(-4a/p) = -(q/p) = (p/q) \\
&= (4a/q) = (4/q)(a/q) = (a/q).
\end{aligned}
$$

In either case, $(a/p) = (q/p)$.

11. Suppose that there are only a finite number of primes of the form $5k - 1$, say, $p_1 < p_2 < \ldots < p_r$. Consider the odd integer $N = 5(p_r!)^2 - 1$. If all the prime divisors of N were of the form $5k + 1$, then N would be of this type; hence, N has a prime divisor p of the form $5k - 1$. Notice that $p > p_r$; for if $p \leq p_r$, then $p | p_r!$ which leads to $p | 1$, an impossibility. Now $5(p_r!)^2 \equiv 1$ (mod p) or $(5p_r!)^2 \equiv 5$ (mod p), and so $(5/p) = 1$. From Problem 10(a), $p \equiv 9$ or 19 (mod 20), or rather $p \equiv -1$ (mod 5). Thus, there is a prime of the form $5k - 1$ which is larger than p_r, a contradiction.

13. (a) Let p be a prime divisor of $839 = 38^2 - 5 \cdot 11^2$. Clearly, $p \neq 2, 3$ or 5. Since $38^2 \equiv 5 \cdot 11^2$ (mod p), we have $1 = (5 \cdot 11^2/p) = (5/11)$ and so, by Problem 10(a), p is of the form $10k + 1$ or $10k + 9$. The only primes to consider are 11 and 19, for $839 < 29^2$. Because neither 11 nor 19 divides 839, it must be prime.

    (b) Assume that $p > 3$ is a prime divisor of $397 = 20^2 - 3$. Then $20^2 \equiv 3$ (mod p), which implies that $(3/p) = 1$; hence, $p \equiv \pm 1$ (mod 12) by Theorem 9-10. Since $397 < 20^2$, the only primes to test are 11 and 13. Neither of these divides 397, so that 397 is prime.

    Let $p > 3$ be a prime divisor of $733 = 29^2 - 3 \cdot 6^2$. Then $29^2 \equiv 3 \cdot 6^2$ (mod p) and so $1 = (3 \cdot 6^2/p) = (3/p)$. It follows that $p \equiv \pm 1$ (mod 12). Since $733 < 29^2$, the primes to consider are 11, 13 and 23, none of which divides 733. Thus 733 is prime.

15. Consider a prime p of the form $p = 2^{4n+1}$. Note that $n \equiv 1$ or 2 (mod 3). For if $n = 3k$, we get $p = 2^{12k+1} = (2^{4k} + 1)(2^{8k} - 2^{4k} + 1)$ and p is not prime. It follows that $p \equiv 3$ or 5 (mod 7): if $n = 3k + 1$,

$$
p = 2^{4(3k+1)} + 1 = 8^{4k} \cdot 2^4 + 1 \equiv 1 \cdot 2 + 1 \equiv 3 \pmod 7
$$

while if $n = 3k + 2$,

$$
p = 2^{4(3k+2)} + 1 = 8^{4k} \cdot 2^8 + 1 \equiv 1 \cdot 4 + 1 \equiv 5 \pmod 7.
$$

Since $p \equiv 1 \pmod 4$, $(7/p) = (p/7)$. Now if $p \equiv 3 \pmod 7$, then $(p/7) = (3/7) = -1$: when $p \equiv 5 \pmod 7$, $(p/7) = (5/7) = -1$. In any case, $(7/p) = -1$, so that $7^{p-1/2} \equiv -1 \pmod p$.

The order of 7 modulo p divides $p - 1 = 2^{4n}$. If the order were less than $p - 1$, it would be equal to 2^j, where $j < 4n$. Then $7^{2^j} \equiv 1 \pmod p$. Successive squarings would lead to $7^{(p-1)/2} \equiv 1 \pmod p$, implying that $p = 2$, a contradiction. Thus, the order of 7 modulo p is $p - 1$, which makes 7 a primitive root of p.

17. Assume that a is a quadratic residue of $b = p_1 p_2 \cdots p_r$, where $p_1, p_2, \ldots, p_r$ are odd primes. Since $x^2 \equiv a \pmod b$ is solvable, so is $x^2 \equiv a \pmod{p_k}$ for $1 \le k \le r$, whence $(a/p_k) = 1$. Thus

$$(a/b) = (a/p_1)(a/p_2) \cdots (a/p_r) = 1 \cdot 1 \cdots 1 = 1.$$

But the converse is false: $(3/35) = (3/5)(3/7) = (-1)(-1) = 1$, yet 3 is not a quadratic residue of 35.

19. Let $\gcd(a, b) = 1$, $a = p_1 p_2 \cdots p_r$ and $b = q_1 q_2 \cdots q_s$, where the p_i and q_j are odd primes, not necessarily distinct. Using parts (a) and (b) of Problem 18,

$$
\begin{aligned}
(a/b)(b/a) &= \prod_{i=1}^{r}\prod_{j=1}^{s}(p_i/q_j) \cdot \prod_{i=1}^{r}\prod_{j=1}^{s}(q_j/p_i) \\
&= \prod_{i=1}^{r}\prod_{j=1}^{s}(p_i/q_j)(q_j/p_i) \\
&= \prod_{i=1}^{r}\prod_{j=1}^{s}(-1)^{(p_i-1)/2 \cdot (q_j-1)/2} \\
&= (-1)^u,
\end{aligned}
$$

with

$$
\begin{aligned}
u &= \sum_{i=1}^{r}\sum_{j=1}^{s}(p_i - 1)/2 \cdot (q_j - 1)/2 \\
&= \sum_{i=1}^{r}(p_i - 1)/2 \cdot \sum_{j=1}^{s}(q_j - 1)/2.
\end{aligned}
$$

But, as was seen in part (f) of Problem 18,

$$\sum_{i=1}^{r}(p_i - 1)/2 \equiv ((p_1 p_2 \cdots p_r) - 1)/2 = (a - 1)/2 \pmod 2$$

$$\sum_{j=1}^{s}(q_j - 1)/2 \equiv ((q_1 q_2 \cdots q_s) - 1)/2 = (b-1)/2 \pmod 2$$

so that $u \equiv (a-1)/2 \cdot (b-1)/2 \pmod 2$. Thus

$$(a/b)(b/a) = (-1)^u = (-1)^{(a-1)/2 \cdot (b-1)/2}.$$

9.4 Quadratic Congruences with Composite Moduli

1. (a) Since $(-1/5) = 1$, Theorem 9-11 implies that the congruence $x^2 \equiv -1 \pmod{52}$ is solvable. From Theorem 8-12, there are exactly two solutions. These are obtained using the argument of Theorem 9-11. First, solve $x^2 \equiv -1 \pmod 5$ for $x \equiv 2$ and 3 $\pmod 5$. In the case $x \equiv 2 \pmod 5$ we have $2^2 = -1 + 1 \cdot 5$, so look for the unique solution of the linear congruence $4y \equiv -1 \pmod 5$; namely, $y \equiv 1 \pmod 5$. Then $x = 2 + 1 \cdot 5 = 7$ will satisfy $x^2 \equiv -1 \pmod{5^2}$. For $x \equiv 3 \pmod 5$, $3^2 = -1 + 2 \cdot 5$, so solve $6y \equiv -2 \pmod 5$ for $y \equiv 3 \pmod 5$. Then $x^2 \equiv -1 \pmod{52}$ has $x = 3 + 3 \cdot 5 = 18$ as its second incongruent solution.

   (b) The two solutions $x \equiv 7, 18 \pmod{52}$ of $x^2 \equiv -1 \pmod{5^2}$ will lead to solutions of $x^2 \equiv -1 \pmod{5^3}$.

   For $x \equiv 7 \pmod{52}$, $7^2 = -1 + 2 \cdot 5^2$, so solve $14y \equiv -2 \pmod 5$ for $y \equiv 2 \pmod 5$. Then $x = 7 + 2 \cdot 5^2 = 57$ satisfies $x^2 \equiv -1 \pmod{53}$.

   For $x \equiv 18 \pmod{52}$, $18^2 = -1 + 13 \cdot 5^2$ and we consider the linear congruence $36y \equiv -13 \pmod 5$. Its solution $y \equiv 2 \pmod 5$ gives $x = 18 + 2 \cdot 5^2 = 68$ as the other solution of $x^2 = -1 \pmod{5^3}$.

3. To find solutions of $x^2 \equiv 31 \pmod{11^4}$, we first solve $x^2 \equiv 31 \pmod{11^2}$ and then $x^2 \equiv 31 \pmod{11^3}$.

   Now $x \equiv 47, 74 \pmod{11^2}$ satisfy $x^2 \equiv 31 \pmod{11^2}$. For $x \equiv 47 \pmod{11^2}$, $47^2 = 31 + 18 \cdot 11^2$, and $94y \equiv -18 \pmod{11}$ has solution $y \equiv 8 \pmod{11}$; hence $x = 47 + 8 \cdot 11^2 = 1015$ satisfies $x^2 \equiv 31 \pmod{11^3}$.

   Now $1015^2 = 31 + 774 \cdot 11^3$ and $2030y \equiv -774 \pmod{11}$ is satisfied by $y \equiv 3 \pmod{11}$. This implies that $x = 1015 + 3 \cdot 11^3 = 5008$ is a solution of $x^2 \equiv 31 \pmod{11^4}$. The other solution is $x \equiv -5008 \equiv 9633 \pmod{11^4}$.

5. Suppose that x_0 is a solution of the quadratic congruence $x^2 \equiv a \pmod{2^n}$, where a is odd and $n \geq 3$. If x_1 is any other solution of this congruence, then $x_0^2 \equiv x_1^2 \pmod{2^n}$ or

$$(x_0 - x_1)(x_0 + x_1) \equiv 0 \pmod{2^n}.$$

Since x_0 and x_1 are both odd, the foregoing congruence is equivalent to

$$(x_0 - x_1)/2 \cdot (x_0 + x_1)/2 \equiv 0 \pmod{2^{n-2}}.$$

Now $(x_0 - x_1)/2 + (x_0 + x_1)/2 = x_0$, and therefore exactly one of $(x_0 - x_1)/2$ and $(x_0 + x_1)/2$ is odd. Consequently, the other is divisible by 2^{n-1}. This implies that one of the congruences $x_0 \pm x_1 \equiv 0 \pmod{2^{n-1}}$ holds; that is, $x_1 \equiv \pm x_0 \pmod{2^{n-1}}$ or $x_1 = \pm x_0 + k2^{n-1}$ for some k. Considering the cases where k may be even or odd, we are led to just four incongruent solutions to $x^2 \equiv a \pmod{2^n}$, namely,

$$x \equiv x_0, \ -x_0, \ x_0 + 2^{n-1}, \ -x_0 + 2^{n-1} \pmod{2^n}.$$

7. (a) By Theorem 9-13, the quadratic congruence $x^2 \equiv a \pmod{2^4}$ has a solution provided that $a \equiv 1 \pmod{8}$; that is, when $a \equiv 1$ or $9 \pmod{2^4}$. For the congruence $x^2 \equiv 1 \pmod{2^4}$, the four solutions are

$$1, \ -1 \equiv 14, \ 1 + 2^3 \equiv 9, \ -1 + 2^3 \equiv 7 \pmod{2^4}.$$

For the congruence $x^2 \equiv 9 \pmod{2^4}$, the solutions are

$$3, \ -3 \equiv 13, \ 3 + 2^3 \equiv 11, \ -3 + 2^3 \equiv 5 \pmod{2^4}.$$

(b) Theorem 9-13 implies that the quadratic congruence $x^2 \equiv a \pmod{2^5}$ is solvable when $a \equiv 1 \pmod{8}$; that is, when $a \equiv 1, 9, 17, 25 \pmod{2^5}$:

For $x^2 \equiv 1 \pmod{2^5}$, the solutions are

$$x \equiv 1, \ 15, \ 17, \ 31 \pmod{2^5}.$$

For $x^2 \equiv 9 \pmod{2^5}$, the solutions are

$$x \equiv 3, \ 13, 19, 29 \pmod{2^5}.$$

For $x^2 \equiv 17 \pmod{2^5}$, the solutions are

$$x \equiv 7,\ 9,\ 23,\ 25 \pmod{2^5}.$$

For $x^2 \equiv 25 \pmod{2^5}$, the solutions are

$$x \equiv 5,\ 11,\ 21,\ 27 \pmod{2^5}.$$

(c) The quadratic congruence $x^2 \equiv a \pmod{2^6}$ has a solution when
$a \equiv 1 \pmod 8$; that is, when $a \equiv 1, 9, 17, 25, 33, 41, 49, 57$
$\pmod{2^6}$:

$x^2 \equiv 1 \pmod{2^6}$ has solutions

$$x \equiv 1,\ 31,\ 33,\ 63 \pmod{2^6}$$

$x^2 \equiv 9 \pmod{2^6}$ has solutions

$$x \equiv 3,\ 29,\ 35,\ 61 \pmod{2^6}$$

$x^2 \equiv 17 \pmod{2^6}$ has solutions

$$x \equiv 9,\ 23,\ 41,\ 55 \pmod{2^6}$$

$x^2 \equiv 25 \pmod{2^6}$ has solutions

$$x \equiv 5,\ 27,\ 37,\ 59 \pmod{2^6}$$

$x^2 \equiv 33 \pmod{2^6}$ has solutions

$$x \equiv 15,\ 17,\ 47,\ 49 \pmod{2^6}$$

$x^2 \equiv 41 \pmod{2^6}$ has solutions

$$x \equiv 13,\ 19,\ 45,\ 51 \pmod{2^6}$$

$x^2 \equiv 49 \pmod{2^6}$ has solutions

$$x \equiv 7,\ 25,\ 39,\ 57 \pmod{2^6}$$

$x^2 \equiv 57 \pmod{2^6}$ has solutions

$$x \equiv 11,\ 21,\ 43,\ 53 \pmod{2^6}$$

9. (a) The quadratic congruence $x^2 \equiv 3 \pmod{11^2 \cdot 23^2}$ will have four solutions, since $x^2 \equiv 3 \pmod{11^2}$ and $x^2 \equiv 3 \pmod{23^2}$ have two solutions each. The congruence $x^2 \equiv 9 \pmod{2^2 \cdot 3 \cdot 5^2}$ will have $4 \cdot 1 \cdot 2 = 8$ solutions, since $x^2 \equiv 9 \pmod{2^3}$ has four solutions, $x^2 \equiv 9 \pmod{3}$ has one solution and $x^2 \equiv 9 \pmod{5^2}$ has two solutions.

(b) To obtain the solutions to $x^2 \equiv 9 \pmod{2^3 \cdot 3 \cdot 5^2}$, first solve each of the congruences $x^2 \equiv 9 \pmod{2^3}$, $x^2 \equiv 9 \pmod{3}$ and $x^2 \equiv 9 \pmod{5^2}$. They have, respectively, the solutions $x \equiv 1, 3, 5, 7 \pmod{2^3}$, $x \equiv 0 \pmod{3}$ and $x \equiv 3, 22 \pmod{5^2}$. Now, using the Chinese Remainder Theorem, solve the following eight systems of linear congruences:

$$
\begin{array}{lll}
x \equiv 1 \pmod{2^3} & x \equiv 1 \pmod{2^3} & x \equiv 3 \pmod{2^3} \\
x \equiv 3 \pmod{5^2} & x \equiv 22 \pmod{5^2} & x \equiv 3 \pmod{5^2} \\
x \equiv 0 \pmod{3} & x \equiv 0 \pmod{3} & x \equiv 0 \pmod{3}
\end{array}
$$

$$
\begin{array}{lll}
x \equiv 3 \pmod{2^3} & x \equiv 5 \pmod{2^3} & x \equiv 5 \pmod{2^3} \\
x \equiv 22 \pmod{5^2} & x \equiv 3 \pmod{5^2} & x \equiv 22 \pmod{5^2} \\
x \equiv 0 \pmod{3} & x \equiv 0 \pmod{3} & x \equiv 0 \pmod{3}
\end{array}
$$

$$
\begin{array}{ll}
x \equiv 7 \pmod{2^3} & x \equiv 7 \pmod{2^3} \\
x \equiv 3 \pmod{5^2} & x \equiv 22 \pmod{5^2} \\
x \equiv 0 \pmod{3} & x \equiv 0 \pmod{3}
\end{array}
$$

The respective solutions are $x \equiv 153, 297, 3, 147, 453, 597, 303, 447 \pmod{2^3 \cdot 3 \cdot 5^2}$.

Chapter 10

Introduction to Cryptography

10.1 From Caesar Cipher to Public Key Cryptography

1. RETURN HOME

3. (a) Suppose that $C \equiv aP + b \pmod{26}$, where $\gcd(a, 26) = 1$. Then the linear congruence $ax \equiv 1 \pmod{26}$ has a unique solution a'. Hence $a'(C - b) \equiv (a'a)P \equiv P \pmod{26}$.

   (b) For the linear cipher $C \equiv 5P + 11 \pmod{26}$, the plaintext message NUMBER THEORY IS EASY encrypts as

   $$\text{YHTQFS \quad CUFDSB \quad ZX \quad FLXB.}$$

   (c) If the linear cipher $C \equiv 3P + 7 \pmod{26}$ is used, then by part (a), $P \equiv 9(C - 7) \pmod{26}$. Thus the cipher message

   $$\text{RZQTGU \quad HOZTKGH \quad DJ \quad KTMMTG}$$

   decrypts as MODERN ALGEBRA IS BETTER.

5. (a) The plaintext

   $$\text{HAVE \quad A \quad NICE \quad TRIP}$$
   $$\text{MATH \quad M \quad ATHM \quad ATHM}$$

   is numerically equivalent to

81

| 07 | 00 | 21 | 04 | | 00 | | 13 | 08 | 02 | 04 | | 19 | 17 | 08 | 15 |
| 12 | 00 | 19 | 07 | | 12 | | 00 | 19 | 07 | 12 | | 00 | 19 | 07 | 12 . |

Adding the columns modulo 26 produces the ciphertext

| 19 | 00 | 14 | 11 | | 12 | | 13 | 01 | 09 | 16 | | 19 | 10 | 15 | 01 |

or, translated into letters,

$$\text{TAOL \quad M \quad NBJQ \quad TKPB.}$$

(b) Since the keyword YES is numerically equivalent to 24 04 18, each 3-letter plaintext block is enciphered by means of the congruences

$$
\begin{aligned}
C_1 &\equiv P_1 + 24 \pmod{26} \\
C_2 &\equiv P_2 + 4 \pmod{26} \\
C_3 &\equiv P_3 + 18 \pmod{26}.
\end{aligned}
$$

The corresponding deciphering congruences are

$$
\begin{aligned}
P_1 &\equiv C_1 - 24 \equiv C_1 + 2 \pmod{26} \\
P_2 &\equiv C_2 - 4 \equiv C_2 + 22 \pmod{26} \\
P_3 &\equiv C_3 - 18 \equiv C_3 + 8 \pmod{26}.
\end{aligned}
$$

Thus the ciphertext

$$\text{BS FMX KFSGR JAPWL}$$

reads in plaintext as

$$\text{DO NOT SHOOT FIRST.}$$

7. (a) We first convert the message GIVE THEM TIME into its numerical equivalent

| 06 | 08 | 21 | 04 | | 19 | 07 | 04 | 12 | | 19 | 08 | 12 | 04. |

Each block of two numbers is now Hill-enciphered to give the following (all modulo 26):

$$5 \cdot 6 + 2 \cdot 8 \equiv 46 \equiv 20, \qquad 3 \cdot 6 + 4 \cdot 8 \equiv 50 \equiv 24;$$
$$5 \cdot 21 + 2 \cdot 4 \equiv 113 \equiv 9, \qquad 3 \cdot 21 + 4 \cdot 4 \equiv 79 \equiv 1;$$
$$5 \cdot 19 + 2 \cdot 7 \equiv 109 \equiv 5, \qquad 3 \cdot 19 + 4 \cdot 7 \equiv 85 \equiv 7;$$
$$5 \cdot 4 + 2 \cdot 12 \equiv 44 \equiv 18, \qquad 3 \cdot 4 + 4 \cdot 12 \equiv 60 \equiv 8;$$
$$5 \cdot 19 + 2 \cdot 8 \equiv 111 \equiv 7, \qquad 3 \cdot 19 + 4 \cdot 8 \equiv 89 \equiv 11;$$
$$5 \cdot 12 + 2 \cdot 4 \equiv 68 \equiv 16, \qquad 3 \cdot 12 + 4 \cdot 4 \equiv 52 \equiv 0.$$

Alphabetically, the ciphertext

20 24 09 01   05 07 18 08   07 11 16 00

reads

$$\text{UYJB FHSI HLQA.}$$

(b) The ciphertext ALXWU VADCOJO is expressed numerically as

00 11 23 22 20   21 00 03 02 14 09 14

Applying the relations

$$P_1 \equiv -8C_1 + 11C_2 \pmod{26}$$
$$P_2 \equiv 3C_1 - 4C_2 \pmod{26}$$

to successive ciphertext pairs, we get

17 08 06 07 19   02 07 14 08 02 04 23

which in letters is

$$\text{RIGHT CHOICEX.}$$

9. (a) The message GO SOX appears in Baudot code as

$$01011000111010000011110111.$$

(b) If the keyword is

$$01110101111010110010100100,$$

then adding corresponding entries in these displays modulo 2 yields the ciphertext

$$001011010000001100100101$$

or, stated in letters,

HS TZM.

11. If $n = pq$, then

$$\phi(n) = (p-1)(q-1) = pq - p - q + 1 = n - p = q + 1$$

Hence $p + q = n - \phi(n) + 1$. Also,

$$(p-q)^2 = (p+q)^2 - 4pq = (p+q)^2 - 4n.$$

In the case where $n = 274279$ and $\phi(n) = 272376$, it follows that

$$
\begin{aligned}
p + q &= 274279 - 272376 + 1 = 1904 \\
p - q &= \sqrt{1904^2 - 4 \cdot 274279} = 1590.
\end{aligned}
$$

The solution of this system of equations is $p = 1747$ and $q = 157$.

13. For the message GOLD MEDAL, the plaintext number is

$$M = 06141103991204030011.$$

Using the encryption key $(2419, 3)$, the message may be broken into blocks of length 3. Then, modulo 2419,

$$
\begin{array}{llll}
061^3 \equiv 2014, & 411^3 \equiv 1231, & 039^3 \equiv 1263, & 912^3 \equiv 508, \\
040^3 \equiv 1106, & 300^3 \equiv 1541, & 11^3 \equiv 1331,
\end{array}
$$

so that the ciphertext message is

2014 1231 1263 0508 1106 1541 1331.

15. For the cryptosystem with key $(2419, 211)$, the recovery exponent satisfies $211j \equiv 1 \pmod{\phi(2419)}$ or $211j \equiv 1 \pmod{2320}$. The solution, modulo 2320, is

$$
\begin{aligned}
j &\equiv 211^{\phi(2320)-1} \\
&\equiv 211^{895} \\
&\equiv (211^{28})^{31} \cdot 211^{27} \equiv 1^{31} \cdot 211^{27} \equiv 11.
\end{aligned}
$$

Given the ciphertext message

$$1369 \quad 1436 \quad 0119 \quad 0385 \quad 0434 \quad 1580 \quad 0690$$

and working modulo 2419,

$$1369^{11} \equiv 180, \quad 1436^{11} \equiv 411, \quad 0119^{11} \equiv 119, \quad 0385^{11} \equiv 1908$$
$$0434^{11} \equiv 071, \quad 1580^{11} \equiv 417, \quad 0690^{11} \equiv 19.$$

The plaintext is $M = 18041111991807141719$ or

$$\text{SELL SHORT.}$$

10.2   The Knapsack Cryptosystem

1. Given the problem

$$21 = 2x_1 + 3x_2 + 5x_3 + 7x_4 + 9x_5 + 11x_6,$$

11 does not fill the knapsack and may be returned; that is, let $x_6 = 1$. Rewrite the problem as

$$10 = 2x_1 + 3x_2 + 5x_3 + 7x_4 + 9x_5.$$

Here 9 cannot be kept, for the addition of it to any other coefficient yields a sum larger than 10; thus, $x_5 = 0$. The equation

$$10 = 2x_1 + 3x_2 + 5x_3 + 7x_4$$

can be satisfied by taking either $x_1 = x_2 = x_3 = 1$, $x_4 = 0$ or by $x_2 = x_4 = 1$, $x_1 = x_3 = 0$. This gives us two solutions to the original problem:

$$x_1 = x_2 = x_3 = x_6 = 1, \quad x_1 = x_3 = x_5 = 0;$$
$$x_2 = x_4 = x_6 = 1, \quad x_1 = x_3 = x_5 = 0.$$

3. (a) Given the knapsack problem

$$118 = 4x_1 + 5x_2 + 10x_3 + 20x_4 + 41x_5 + 99x_6,$$

let $x_6 = 1$, so as to retain 99. In the reduced problem,

$$19 = 4x_1 + 5x_2 + 10x_3 + 20x_4 + 41x_5$$

take $x_4 = x_5 = 0$ and consider the equation

$$19 = 4x_1 + 5x_2 + 10x_3.$$

This is satisfied by $x_1 = x_2 = x_3 = 1$.

(b) For the problem

$$51 = 3x_1 + 5x_2 + 9x_3 + 18x_4 + 37x_5,$$

letting $x_5 = 1$ leads to the reduced problem

$$14 = 3x_1 + 5x_2 + 9x_3 + 18x_4.$$

Here, $x_4 = 0$ and the resulting equation

$$14 = 3x_1 + 5x_2 + 9x_3$$

has $x_1 = 0$, $x_2 = x_3 = 1$ as its solution.

(c) The equation

$$54 = x_1 + 2x_2 + 5x_3 + 9x_4 + 18x_5 + 40x_6$$

indicates that $x_6 = 1$. In the reduced problem

$$14 = x_1 + 2x_2 + 5x_3 + 9x_4 + 18x_5$$

set $x_5 = 0$ and consider

$$14 = x_1 + 2x_2 + 5x_3 + 9x_4.$$

For this equation, put $x_1 = x_2 = 0$ and $x_3 = x_4 = 1$.

5. Note that $33 \cdot 47 \equiv 1 \pmod{50}$. Since

$$47 \cdot 49 = 2303 \equiv 3 \pmod{50}, \qquad 47 \cdot 32 = 1504 \equiv 4 \pmod{50},$$
$$47 \cdot 30 = 1410 \equiv 10 \pmod{50}, \qquad 447 \cdot 42 = 2021 \equiv 21 \pmod{50},$$

the original superincreasing sequence is 3, 4, 10, 21.

7. (a) Given the superincreasing sequence 2, 3, 7, 13, 27, multiply each term by 7 and work modulo 40 to obtain

$$7 \cdot 2 \equiv 14, \ \ 7 \cdot 3 \equiv 21, \ \ 7 \cdot 7 \equiv 9, \ \ 7 \cdot 13 \equiv 11, \ \ 7 \cdot 27 \equiv 29.$$

(b) Each letter of the message SEND MONEY is converted to its binary equivalent to produce the numerical representation

$$10010 \quad 00100 \quad 01101 \quad 00011,$$
$$01100 \quad 01110 \quad 01101 \quad 00100 \quad 11000.$$

Using the terms of the listed public key as multipliers, the blocks of digits become

$$25 = 14 + 11 \quad 21 = 21 \qquad 59 = 21 + 9 + 29 \quad 40 = 11 + 29,$$

$$30 = 21 + 9 \quad 41 = 21 + 9 + 11 \quad 59 = 21 + 9 + 29 \quad 9 = 9$$
$$35 = 14 + 21.$$

Thus the encrypted message is

$$25 \ 21 \ 59 \ 40 \ 30 \ 41 \ 59 \ 9 \ 35.$$

10.3 An Application of Primitive Roots to Cryptography

1. (a) The message REPLY TODAY has as its digital equivalent

$$M = 17041511241914030024.$$

For the random integer $j = 13$, $5^{13} \equiv 43 \pmod{47}$ and $10^{13} \equiv 38 \pmod{47}$. Thus each two-digit block in M is multiplied by 38 and reduced modulo 47 to produce the ciphertext

$$(43, 35) \quad (43, 11) \quad (43, 06) \quad (43, 42) \quad (43, 19)$$
$$(43, 17) \quad (43, 15) \quad (43, 20) \quad (43, 00) \quad (43, 19)$$

(b) In order to recover the message M, multiply each second entry in the ordered pairs by $43^{46-19} = 43^{27} \equiv 26 \pmod{47}$. For instance $26 \cdot 35 \equiv 17 \pmod{47}$ yields the first letter R of the plaintext.

3. To encrypt the message 131419131422, which is to be sent to a person with public key $(37, 2, 18)$ and private key $k = 17$, break the message into successive four-digit blocks and then encipher each two-digit sub-block. Taking $j = 15$ as the random enciphering integer for the block 1314, we have $2^{15} \equiv 23 \pmod{37}$ and $8^{15} \equiv 30 \pmod{37}$. Because $8 \cdot 13 \equiv 17 \pmod{37}$ and $8 \cdot 14 \equiv 01 \pmod{27}$, the corresponding ElGamal ciphertext is $(23, 30) \ (23, 01)$.

If $j = 28$ is the random integer for the block 1913, then $2^{28} \equiv 12 \pmod{37}$ and $18^{28} \equiv 34 \pmod{37}$. The congruences $34 \cdot 19 \equiv 17$

(mod 37) and $34 \cdot 13 \equiv 35 \pmod{37}$ imply that the ciphertext in this case is $(12, 17)\ (12, 35)$.

When $j = 11$ is the random integer associated with the block 1422, it follows that $2^{11} \equiv 13 \pmod{37}$ and $18^{11} \equiv 17 \pmod{37}$; hence, $17 \cdot 14 \equiv 16 \pmod{37}$ and $17 \cdot 22 \equiv 04 \pmod{37}$. The ciphertext here is $(13, 16)\ (13, 04)$. The entire message is encrypted as:

$$(23, 30)\ (23, 01)\ (12, 17)\ (12, 35)\ (13, 16)\ (13, 04).$$

5.   (a) The signature (c, d) for this message is found by computing

$$c \equiv 3^{13} \equiv 24 \pmod{31}$$

and the solution $d \equiv 2 \pmod{30}$ of the congruence

$$13d \equiv 14 - 17 \cdot 24 \pmod{30}.$$

     (b) The validity is confirmed from the equality of

$$V_1 \equiv 22^{24} \cdot 24^2 \equiv 10 \pmod{31}$$

and

$$V_2 \equiv 3^{14} \equiv 10 \pmod{31}.$$

Chapter 11

Numbers of Special Form

11.2 Perfect Numbers

1. If $n = 2^{10}(2^{11} - 1) = 2^{10} \cdot 23 \cdot 89$, then

$$
\begin{aligned}
\sigma(n) &= \sigma(2^{10})\sigma(23)\sigma(89) \\
&= (2^{11} - 1)24 \cdot 90 = 2160(2^{11} - 1) \\
&\neq 2 \cdot 2^{10}(2^{11} - 1) = 2n.
\end{aligned}
$$

3. Suppose n is a perfect number. If $d|n$, then $n = dd'$ for some d', where $d'|n$. Hence

$$
\sum_{d|n} 1/d = \sum_{d'|n} d'/n = 1/n \sum_{d'|n} d' = (1/n)2n = 2.
$$

5. Let $n = 2^{k-1}(2^k - 1)$ be an even perfect number. By Problem l(a) of Section 1.1,

$$
\begin{aligned}
1 + 2 + 3 + \cdots + (2^k - 1) &= \frac{(2^k - 1)((2^k - 1) + 1)}{2} \\
&= (2^k - 1)2^{k-1} = n.
\end{aligned}
$$

Also, since $2^k - 1$ is prime,

$$
\begin{aligned}
\phi(n) &= \phi(2^{k-1})\phi(2^k - 1) = (2^{k-1} - 2^{k-2})(2^k - 2) \\
&= 2^{k-2}(2 - 1)(2^k - 2) = 2^{k-1}(2^{k-1} - 1).
\end{aligned}
$$

9.   (a) If $n = 523,776 = 2^9 \cdot 3 \cdot 11 \cdot 31$, then

$$
\begin{aligned}
\sigma(n) &= \sigma(2^9)\sigma(3)\sigma(11)\sigma(31) \\
&= (2^{10} - 1)4 \cdot 12 \cdot 32 \\
&= (3 \cdot 11 \cdot 31)4 \cdot 12 \cdot 32 = 3(2^9 \cdot 3 \cdot 11 \cdot 31) = 3n.
\end{aligned}
$$

If $n = 30,240 = 2^5 \cdot 3^3 \cdot 5 \cdot 7$, then

$$
\begin{aligned}
\sigma(n) &= \sigma(2^5)\sigma(3^3)\sigma(5)\sigma(7) \\
&= (2^6 - 1)40 \cdot 6 \cdot 8 \\
&= (9 \cdot 7)40 \cdot 6 \cdot 8 = 4(2^5 \cdot 3^3 \cdot 5 \cdot 7) = 4n.
\end{aligned}
$$

If $n = 114,182,439,040 = 2^7 \cdot 3^4 \cdot 5 \cdot 7 \cdot 11^2 \cdot 17 \cdot 19$, then

$$
\begin{aligned}
\sigma(n) &= \sigma(2^7)\sigma(3^4)\sigma(5)\sigma(7)\sigma(11^2)\sigma(17)\sigma'(19) \\
&= (2^8 - 1)121 \cdot 6 \cdot 8 \cdot 133 \cdot 18 \cdot 20 \\
&= (3 \cdot 5 \cdot 17)11^2 \cdot 6 \cdot 8 \cdot 7 \cdot 19 \cdot 18 \cdot 20 \\
&= 5(2^7 \cdot 3^4 \cdot 5 \cdot 7 \cdot 11^2 \cdot 17 \cdot 19) = 5n.
\end{aligned}
$$

11. If $n^2 = \prod_{d \mid n} d = n^{\tau(n)/2}$, then $\tau(n) = 4$. This means that either $n = p^3$ or $n = pq$, where p and q are distinct primes.

13. Suppose $n = 2^{k-1}(2^k - 1)$ is an even perfect number, so that $2^k - 1$ is prime. Then

$$
\begin{aligned}
\sigma(n^2) + 1 &= \sigma(2^{2k-2})\sigma((2^k - 1)^2) + 1 \\
&= (2^{2k-1} - 1)((2^k - 1)^2 + (2^k - 1) + 1) + 1 \\
&= (2^{2k-1} - 1)(2^{2k} - 2^{k+1} + 2^k + 1) + 1 \\
&= 2^k M,
\end{aligned}
$$

where $M = 2^{3k-1} - 2^{2k} + 2^{2k-1} + 2^{k-1} - 2^k + 1$.

15. Assume that n is a perfect number. Using Problem 8 of Section 6.1,

$$
H(n) = n\tau(n)/\sigma(n) = n\tau(n)/2n = \tau(n)/2.
$$

Since n is a perfect number, n cannot be a square by Problem 2(b). Thus, according to Problem 7(a) of section 6.1, $\tau(n)$ is an even integer; hence $H(n)$ is an integer.

17. Notice that

$$2^{k-1} + 2^k + 2^{k+1} + \cdots + 2^{2k-2}$$
$$= 2^{k-1}(1 + 2 + 2^2 + \cdots + 2^{k-1})$$
$$= 2^{k-1}(2^k - 1),$$

which will be a perfect number provided that $2k - 1$ is prime.

19. Suppose that $n = n_1 n_2 \cdots n_r$, where $n_i = 2^{k_i - 1}(2^{k_i} - 1)$ is an even perfect number. By Problem 5, $\phi(n_i) = 2^{k_i - 1}(2^{k_i - 1} - 1)$. Then

$$\phi(n_1 n_2 \cdots n_r) = \phi(2^{k_1 + k_2 + \cdots + k_r - r}(2^{k_1} - 1) \cdots (2^{k_r} - 1))$$
$$= \phi(2^{k_1 + k_2 + \cdots + k_r - r})\phi(2^{k_1} - 1) \cdots \phi(2^{k_r} - 1)$$
$$= 2^{k_1 + k_2 + \cdots + k_r - r - 1}(2^{k_1} - 2) \cdots (2^{k_r} - 2)$$
$$= 2^{k_1 - 1} \cdots 2^{k_r - 1} \cdot 2^{r-1}(2^{k_1 - 1} - 1) \cdots (2^{k_r - 1} - 1)$$
$$= 2^{r-1}\phi(n_1)\phi(n_2) \cdots \phi(n_r).$$

11.3 Mersenne Primes and Amicable Numbers

1. By Theorem 11-5, the prime divisors of $M_{13} = 2^{13} - 1$ are of the form $26k + 1$. Since $\sqrt{M_{13}} < 91$, the only candidates are 53 and 79, neither of which divides M_{13}: hence, M_{13} is prime.

3. The prime divisors of $M_{29} = 2^{29} - 1$ are of the form $58k + 1$. Of these primes, the smallest which is congruent to ± 1 modulo 8 is 233. Also, $M_{29} = 233 \cdot 2304167$.

5. Suppose n is an even perfect number: say, $n = 2^{k-1}(2^k - 1)$, where $2^k - 1$ is prime. Let $1 < d < n$ and $d | n$. If $d = 2^j$, for $0 < j \le k - 1$, then

$$\sigma(d) = \sigma(2^j) = (1 + 2 + \cdots + 2^{j-1}) + 2^j < 2^j + 2^j = 2d.$$

If $\sigma(d) = 2^j(2^k - 1)$, for $0 \le j < k - 1$, then

$$\sigma(d) = \sigma(2^j)\sigma(2^k - 1) = (2^{j+1} - 1)2^k$$
$$= 2 \cdot 2^j(2^k - 2^{k-1-j})$$
$$< 2 \cdot 2^j(2^k - 1) = 2d.$$

In either case, d is deficient.

7. (a) For $220 = 2^2 \cdot 5 \cdot 11$ and $284 = 2^2 \cdot 71$,

$$\begin{aligned}
\sigma(220) - 220 &= 7 \cdot 6 \cdot 12 - 220 = 504 - 220 = 284, \\
\sigma(284) - 284 &= 7 \cdot 72 - 284 = 504 - 284 = 220.
\end{aligned}$$

(b) For $17296 = 2^4 \cdot 23 \cdot 47$ and $18416 = 2^4 \cdot 1151$,

$$\begin{aligned}
\sigma(17296) - 17296 &= 31 \cdot 24 \cdot 48 - 17296 \\
&= 35712 - 17296 = 18416, \\
\sigma(18416) - 18416 &= 31 \cdot 1152 - 18416 \\
&= 35712 - 18416 = 17296.
\end{aligned}$$

(c) For $9363584 = 2^7 \cdot 191 \cdot 383$ and $9437056 = 2^7 \cdot 73727$,

$$\begin{aligned}
\sigma(9363584) \ - \ & 9363584 = 225 \cdot 192 \cdot 384 - 9363584 \\
= \ & 18800640 - 9363584 = 9437056, \\
\sigma(9437056) \ - \ & 9437056 = 255 \cdot 73728 - 9437056 \\
= \ & 18800640 - 9437056 = 9363584.
\end{aligned}$$

9. (a) Suppose that n and the prime p are an amicable pair, hence

$$\sigma(n) = n + p = \sigma(p).$$

The condition $\sigma(p) = n + p$ implies that $n = 1$; but then the condition $\sigma(n) = n + p$, with $n = 1$, yields the contradiction $p = 0$.

(b) If m and n are an amicable pair with $n < m$, then

$$\sigma(m) = m + n < m + m = 2m,$$

which makes m deficient.

(c) Suppose that m and n are an amicable pair, where m is even and n is odd. Then $\sigma(m) = m + n = \sigma(n)$, so that $\sigma(m)$ and $\sigma(n)$ are both odd. Let $n = p_1^{k_1} p_2^{k_2} \cdots p_r^{k_r}$ with p_i an odd prime; then

$$\sigma(n) = (1 + p_1 + \cdots + p_1^{k_1}) \cdots (1 + p_r + \cdots + p_r^{k_r}).$$

Since each factor $(1 + p_i + \cdots + p_i^{k_i})$ is an odd integer, it follows that $k_i = 2j_i$ for all i. This means that n is a square, by Problem 15 of section 3.1.

11. If $p = 3 \cdot 2^{n-1} - 1$, $q = 3 \cdot 2^n - 1$ and $r = 9 \cdot 2^{2n-1} - 1$, for $n \geq 2$, then

$$
\begin{aligned}
\sigma(2^n pq) &= (2^{n+1} - 1)(p+1)(q+1) \\
&= (2^{n+1} - 1)(3 \cdot 2^{n-1})(3 \cdot 2^n) \\
&= (2^{n+1} - 1)9 \cdot r^{2n-1}, \quad \text{and} \\
\sigma(2^n r) &= (2^{n+1} - 1)(r+1) \\
&= (2^{n+1} - 1)9 \cdot 2^{2n-1}.
\end{aligned}
$$

Thus, $\sigma(2^n pq) = (2^{n+1} - 1)9.2^{2n-1} = \sigma(2^n r)$, where

$$
(2^{n+1} - 1)9.2^{n-1} = 2^n pq + 2^n r.
$$

13. For the integers $14288 = 2^4 \cdot 19 \cdot 47$, $15472 = 2^4 \cdot 967$, $14536 = 2^3 \cdot 23 \cdot 79$, $14264 = 2^3 \cdot 1783$ and $12496 = 2^4 \cdot 11 \cdot 71$, we have

$$
\begin{aligned}
\sigma(14288) - 14288 &= 31 \cdot 20 \cdot 48 - 14288 \\
&= 29760 - 14288 = 15472, \\
\sigma(15472) - 15472 &= 31 \cdot 968 - 15472 \\
&= 30008 - 15472 = 14536, \\
\sigma(14536) - 14536 &= 15 \cdot 24 \cdot 80 - 14536 \\
&= 28800 - 14536 = 14264; \\
\sigma(14264) - 14264 &= 15 \cdot 1784 - 14264 \\
&= 26760 - 14264 = 12496, \quad \text{and} \\
\sigma(12496) - 12496 &= 31 \cdot 12 \cdot 72 - 12496 \\
&= 26784 - 12496 = 14288.
\end{aligned}
$$

Thus the given integers form a sociable chain.

15. Suppose there is an odd perfect number n of the form $n = p^k q^j$, where p and q are distinct odd primes. Then

$$
\begin{aligned}
2n = \sigma(n) &= \frac{p^{k+1} - 1}{p - 1} \cdot \frac{q^{j+1} - 1}{q - 1} \\
&= \frac{p^{k+1}}{p - 1} \cdot \frac{q^{j+1}}{q - 1} \\
&= \frac{npq}{(p-1)(q-1)} \\
&= \frac{n}{(1 - 1/p)(1 - 1/q)}.
\end{aligned}
$$

Since $p \geq 3$ and $q \geq 5$, this inequality leads to

$$2 \frac{1}{(1 - 1/p)(1 - 1/q)} \leq \frac{3 \cdot 5}{2 \cdot 4} = 15/8,$$

which is a contradiction. Thus an odd perfect number has at least three distinct prime factors.

11.4   Fermat Numbers

1. Raising the congruence $5 \cdot 2^7 \equiv -1 \pmod{641}$ to the fourth power gives

$$1 \equiv 5^4 \cdot 2^{28} \equiv 625 \cdot 2^{28} \equiv -16 \cdot 2^{28} \equiv -2^{32} \pmod{641}.$$

Hence, $F_5 = 2^{32} + 1 \equiv 0 \pmod{641}$.

3. (a) Since $2^{2j} \equiv 4^j \equiv 1 \pmod 3$, $2^{2j} = 3k + 1$ for some k. Now

$$\begin{aligned}
2^{2^{2j+1}} + 3 &= 2^{2 \cdot 2^{2j}} + 3 = 4^{3k+1} + 3 \\
&= 4(4^3)^k + 3 \equiv 4.1^k + 3 \equiv 7 \equiv 0 \pmod 7.
\end{aligned}$$

Thus, when n is odd, $7 | 2^{2^n} + 3$.

   (b) For $n > 0$,

$$2^{2^n} + 5 = (2^2)^{2^{n-1}} + 5 \equiv 1^{2^{n-1}} + 5 \equiv 6 \equiv 0 \pmod 3,$$

hence $3 | 2^{2^n} + 5$.

5. First, we prove by induction that $2^{2^n} \equiv 6 \pmod{10}$ for $n \geq 2$. When $n = 2$, $2^{2^2} = 2^4 = 16 \equiv 6 \pmod{10}$. Assuming that $2^{2^k} \equiv 6 \pmod{10}$, then

$$2^{2^{k+1}} = (2^{2^k})^2 \equiv 6^2 \equiv 36 \equiv 6 \pmod{10}.$$

Thus, whenever the asserted congruence holds for k, it also holds for $k + 1$.

For $n \geq 2$, this implies that

$$F_n = 2^{2^n} + 1 \equiv 6 + 1 \equiv 7 \pmod{10}.$$

7. Notice that

$$
\begin{aligned}
2^{58} + 1 &= 4(2^{14})4 + 1 \\
&= [2(2^{14})^2 - 2(2^{14}) + 1][2(2^{14})^2 + 2(2^{14}) + 1] \\
&= 536838145 \cdot 536903681.
\end{aligned}
$$

9. (a) If n is odd, say $n = 2k + 1$ for some k, then

$$
2^{2k+1} + 1 = 2.4^k + 1 \equiv 2 \cdot 1^k + 1 \equiv 3 \equiv 0 \pmod 3.
$$

(b) Since $2^p + 1 \equiv 0 \pmod q$ leads to $2^{2p} \equiv 1 \pmod q$, the order of 2 modulo q is one of 2, p or $2p$. But if 2 has order p, then $2^p \equiv 1 \pmod q$, which contradicts the hypothesis $2^p \equiv -1 \pmod q$. If the order of 2 is 2, it follows that $2^2 \equiv 1 \pmod q$ or $q = 3$. If the order of 2 is $2p$, then $2p | \phi(q)$ or $2p | q - 1$; hence $q - 1 = 2kp$ for some k, or $q = 2kp + 1$.

(c) If $q | 2^{29} + 1$, where $q > 3$, then q is of the form $q = 58k + 1$. It happens that $59 | 2^{29} + 1$. If $q | 2^{41} + 1$, with $q > 3$, then the prime $q = 82k + 1$ for some k. In fact, $83 | 2^{29} + 1$.

11. From Problem 9(a), if $p | 2^p + 1$ where p is a prime, then either $p = 3$ or $p = 2kp + 1$ for some k. Since the latter equality is impossible, $p = 3$. From Problem 8(b) of Section 8.1, if $p | 2^p - 1$, then $p = 2kp + 1$ for some k; since this is impossible, there is no such p.

13. Since $2^{2^n} \equiv 2^{2^{n-2}} \pmod 9$ for $n \geq 3$, $2^{2^n} + 1 \equiv 2^{2^2} + 1 \equiv 8 \pmod 9$ or $2^{2^n} + 1 \equiv 2^2 + 1 \equiv 5 \pmod 9$ according as n is even or odd.

15. (a) If n is even, then $1^n \equiv 1 \pmod 3$, while if n is odd, then $2^n \equiv 2 \pmod 3$. Thus for a prime $p > 3$ (so that $p - 1 \geq 4$),

$$
\begin{aligned}
\frac{1}{3}(2^p + 1) &= 2^{p-1} - 2^{p-2} + 2^{p-3} - 2^{p-4} + \cdots - 2 + 1 \\
&\equiv (1 - 2) + (1 - 2) + \cdots + (1 - 2) + 1 \pmod 3 \\
&\equiv \left(\frac{p-1}{2}\right)(-1) + 1 \pmod 3 \\
&\equiv \frac{p+3}{2} \pmod 3.
\end{aligned}
$$

Now 3 does not divide $(2^p + 1)/3$; for $(p + 3)/2 \equiv 0 \pmod 3$ implies that $p = 3$, which is impossible.

(b) By Problem 9(b), the prime divisors of $2^p + 1$ are either 3 or of the form $2kp + 1$. Since 3 does not divide $(2^p + 1)/3$, it is either prime or else has prime divisors which look like $2kp+1$. In either case, $(2^p + 1)/3$ has a prime divisor larger than p.

(c) The integer $(2^{19} + 1)/3$ is either prime or has a prime divisor of the form $38k + 1$. Primes of this form which are less than the square root of $(2^{19} + 1)/3$ are 191 and 229, neither of which divides $(2^{19} + 1)/3$.

The integer $(2^{23} + 1)/3$ is either prime or has a prime divisor of the form $46k + 1$. Candidates for prime divisors are 47, 139, 227, 461, 599, 691, 829, 967, 1013, 1151, 1289, 1381, 1427, 1657, and 1933, none of which divides $(2^{23} + 1)/3$. Hence, $(2^{23} + 1)/3$ is a prime number .

17. (a) Suppose that F_n is a Fermat prime where $n \geq 2$. Then Theorem 11-10 implies that $3^{(F_n-1)/2} \equiv -1 \pmod{F_n}$. Hence, 3 is a quadratic nonresidue of F_n by Euler's Criterion.

Using the Quadratic Reciprocity Law, for $n \geq 2$,

$$
\begin{aligned}
(5/F_n) &= (F_n/5) \equiv F_n^2 \pmod 5 \\
&\equiv 2^{2^{n+1}} + 2 \cdot 2^n + 1 \pmod 5 \\
&\equiv 1 + 2 + 1 \equiv -1 \pmod 5.
\end{aligned}
$$

Thus, $(5/F_n) = -1$, making 5 a quadratic nonresidue of F_n. By Problem 15 of Section 9.3, 7 is a primitive root of any prime of the form $2^{4n} + 1$, hence, of any Fermat prime F_n with $n \geq 2$. Therefore, 7 is a quadratic nonresidue of F_n (Problem 5(a), section 9.1).

(b) Let a be a quadratic nonresidue of the Fermat prime F_n, so that $a^{(F_n-1)/2} \equiv -1 \pmod{F_n}$. Now the order of a modulo F_n divides $\phi(F_n) = F_n - 1 = 2^{2^n}$; hence the order of a is equal to 2^k where $k \leq 2^n$. If the order of a is 2^k, with $k < 2^n$, then $a^{2^k} \equiv 1 \pmod{F_n}$. Repeated squaring of this congruence produces $a^{(F_n-1)/2} \equiv 1 \pmod{F_n}$, which is a contradiction. Thus the order of a is $2^{2^n} = F_n - 1$ and a is a primitive root of F_n.

19. For $n \geq 1$, assume that $\gcd(F_n, n) > 1$. Then there exists a prime p satisfying $p|F_n$ and $p|n$. By Theorem 11-11, $p = k \cdot 2^{n+2} + 1$. But then $p \geq 2^{n+2} + 1 > 2^n > n$, which contradicts $p|n$. Thus $\gcd(F_n, n) = 1$.

Chapter 12

Certain Nonlinear Diophantine Equations

12.1 The Equation $x^2 + y^2 = z^2$

1. (b) If $x = 2st = 40$, then there are two possibilities for s and t: when $s = 5$ and $t = 4$, so that $y = s^2 - t^2 = 9$ and $z = s^2 + t^2 = 41$; and $s = 20$ and $t = 1$, so that $y = s^2 - t^2 = 399$ and $z = s^2 + t^2 = 401$. This gives two triples, $(40, 9, 41)$ and $(40, 399, 401)$.

If $x = 2st = 60$, there are four possibilities for s and t:

When $s = 6$, $t = 5$, then $y = s^2 - t^2 = 11$ and $z = s^2 + t^2 = 61$.

When $s = 10$, $t = 1$, then $y = s^2 - t^2 = 91$ and $z = s^2 + t^2 = 109$.

When $s = 15$, $t = 2$, then $y = s^2 - t^2 = 221$ and $z = s^2 + t^2 = 229$.

When $s = 30$, $t = 1$, then $y = s^2 - t^2 = 889$ and $z = s^2 + t^2 = 901$.

Thus there are four triples: $(60, 11, 61)$, $(60, 91, 109)$, $(60, 221, 229)$ and $(60, 899, 901)$.

3. (a) Let $n \not\equiv 2 \pmod 4$. Suppose $n \equiv 0 \pmod 4$, say $n = 4r$. Using Theorem 12-1, $s = 2r$ and $t = 1$ gives the primitive Pythagorean triple

$$x = n, \ y = 4r^2 - 1, z = 4r^2 + 1.$$

If $n \equiv 1 \pmod 4$, say $n = 4r^2 + 1$, then $s = 2r + 1$ and $t = 2r$ gives the primitive Pythagorean triple

$$x = 8r^2 + 4r, \ y = n, \ z = 8r^2 + 4r + 1.$$

If $n \equiv 3 \pmod 4$, say $n = 4r - 1$, then $s = 2r$ and $t = 2r - 1$ gives the primitive Pythagorean triple

$$x = 8r^2 - 4r, \ y = n, \ z = 8r^2 - 4r + 1.$$

(b) Let $n \equiv 2 \pmod 4$. Then $n = 4r + 2 = 2k$, where $k = 2r + 1 \equiv$ 1 or 3 $\pmod 4$. By part (a), there is a primitive Pythagorean triple (x, k, z) having k as a member. Then $(2x, n, 2z)$ is a Pythagorean triple with n as a member.

5. Given $n \geq 1$, let

$$x = 2^{n+1}, y_k = 2^k(2^{2n-2k} - 1), z_k = 2^k(2^{2n-2k} + 1).$$

Then (x, y_k, z_k) is a Pythagorean triple for $k = 0, 1, \ldots, n - 1$:

$$
\begin{aligned}
x^2 + y^2 &= x^{2n+2} + 2^{2k}(2^{4n-4k} - 2^{2n-2k+1} + 1) \\
&= 2^{2n+2} + 2^{4n-2k} - 2^{2n+1} + 2^{2k} \\
&= 2^{2k}(2^{4n-4k} + 2^{2n-2k+1} + 1) = z^2.
\end{aligned}
$$

Thus there are at least n triples with x as their first member.

7. Let the integers $x - d$, x and $x + d$ make up a Pythagorean triple, so that $(x - d)^2 + x^2 = (x + d)^2$. Then $x^2 = 4dx$, or $x = 4d$. Thus the triple is $3d$, $4d$ and $5d$.

9. Consider a primitive Pythagorean triple

$$x = 2st, \ y = s^2 - t^2, \ z = x^2 + t^2.$$

(a) If $z = x + 1$, then $s^2 + t^2 - 2st = (s - t)^2 = 1$. Since $s > t$, $s - t = 1$ and so $s = t + 1$. Thus the triple is

$$x = 2t(t + 1), \ y = 2t + 1, z = 2t(t + 1) + 1.$$

(b) If $z = y + 2$, then $s^2 + t^2 = s^2 - t^2 + 2$ or $2t^2 = 2$. Hence $t = 1$ and the triple is

$$x = 2s, \ y = s^2 - 1, \ z = s^2 + 1.$$

11. Let r be the radius of the circle inscribed in a right triangle, with sides of length a and b, and hypotenuse of length c. The diagram in Theorem 12-2 indicates that $c = (a - r) + (b - r)$, hence $2r = a + b - c$. For the triangle with sides $2n + 1$, $2n^2 + 2n$, $2n^2 + 2n + 1$, we have

$$2r = (2n + 1) + (2n^2 + 2n) - (2n^2 + 2n + 1) = 2n$$

or $r = n$.

13. Let x, $x+1$, z be a Pythagorean triple, so that $2x^2 + 2x = z^2 - 1$. Put $u = z - x - 1$ and $v = x + (1-z)/2$. Then

$$
\begin{aligned}
t_u &= \frac{u(u+1)}{2} = \frac{1}{2}(z - x - 1)(z - x) \\
&= x^2 + x(l - z) + \frac{1}{4}(2z^2 - 2z - (2x^2 + 2x)) \\
&= x^2 + x(1 - z) + \frac{1}{4}(z^2 - 2z + 1) \\
&= (x + \frac{1}{2}(1 - z))^2 = v^2
\end{aligned}
$$

Since there are infinitely many Pythagorean triples of the form x, $x+1$, z, there are infinitely many triangular numbers which are squares.

12.2 Fermat's Last Theorem

1. For $n > 1$, let $x = n(n^2 - 3)$, $y = 3n^2 - 1$, $z = n^2 + 1$, all of which are positive integers. Then

$$
\begin{aligned}
x^2 + y^2 &= n^2(n^4 - 6n^2 + 9) + (9n^4 - 6n^2 + 1) \\
&= n^2(n^4 + 3n^2 + 3) + 1 \\
&= (n^2 + 1)^3 = z^3.
\end{aligned}
$$

3. Let x, y, z be a Pythagorean triple. If x and y are squares, say $x = a^2$, $y = b^2$, then $a^4 + b^4 = z^2$; this is impossible by Theorem 12-3. If x and z are squares, say $x = a^2$, $z = c^2$, then $y^2 = c^4 - a^4$; this is impossible by Theorem 12-4. Thus, no two members of the triple can be squares. If x, y, z are all squares, Theorem 12-3 is again violated.

5. Suppose to the contrary that the system $x^2 + y^2 = z^2$ and $x^2 + 2y^2 = w^2$ has a simultaneous solution in the positive integers. Then

$$
w^2 = x^2 + 2y^2 = z^2 + y^2.
$$

This implies that the equations $z^2 + y^2 = w^2$ and $z^2 - y^2 = x^2$ are simultaneously solvable, contradicting Problem 4(b).

7. Suppose to the contrary that $x^4 - y^4 = 2z^2$ has a solution in the positive integers. Then x and y are both even, or both odd. This

implies that $x^2 + y^2$, $x + y$ and $x - y$ are all even, say $x^2 + y^2 = 2r$, $x + y = 2s$, $x - y = 2t$. Then

$$2z^2 = x^4 - y^4 = 8rst$$

or $z^2 = 4rst$, so that r, s, t are all squares. Put $r = a^2$, $s = b^2$, $t = c^2$. The equation $2(x^2 + y^2) = (x + y)^2 + (x - y)^2$ gives $2(2a^2) = (2b^2)^2 + (2c^2)^2$ or $a^2 = b^4 + c^4$. But the equation $u^4 + v^4 = w^2$ has no solution in the positive integers by Theorem 12-3, a contradiction.

9. Assume to the contrary that the equation $x^4 - 4y^4 = z^2$ has a solution. Then
$$(x^4 - 4y^4)^2 + 16x^4y^4 = z^4 + 16x^4y^4$$

or $(x^4 + 4y^4)^2 = x^4 + (2xy)^4$, so that the equation $u^4 + v^4 = w^2$ has a solution. But this is impossible by Theorem 12-3.

11. Assume $1/x^2 + 1/y^2 = 1/z^2$, where $\gcd(x, y, z) = 1$, has a solution in the positive integers ($x > z$). Then $y^2(x^2 - z^2) = x^2z^2$. Let $d = \gcd(x, z)$, so that $x = ad$, $z = cd$ for some c, d with $\gcd(a, c) = 1$. Then $y^2(a^2 - c^2) = (dac)^2$; hence $y^2|(dac)^2$ or $y|dac$, say $dac = yb$ for some b. It follows that $a^2 - c^2 = b^2$ or $b^2 + c^2 = a^2$, with $\gcd(a, b) = \gcd(b, c) = 1$. By Theorem 12-1,

$$b = 2st, \qquad c = s^2 - t^2, \quad a = s^2 + t^2$$
$$\gcd(s, t) = 1, \quad s > t, \qquad s \equiv t \pmod 2.$$

Since $\gcd(b, ac) = 1$, $dac = yb$ implies that $b|d$, say $d = br$. But then $x = abr$, $y = acr$, $z = bcr$. Since $\gcd(x, y, z) = 1$, this gives $r = 1$. Thus

$$x = ab = 2st(s^2 + t^2), \; y = ac = (s^4 - t^4), \; z = bc = 2st(s^2 - t^2).$$

Chapter 13

Representation of Integers as Sums of Squares

13.2 Sums of Two Squares

3. (a) If n is odd, then $2^n = (2^{(n-1)/2})^2 + (2^{(n-1)/2})^2$; while if n is even, then $2^n = (2^{n/2})^2 + 0^2$, for $n \geq 1$.

    (b) Suppose that $n \equiv 3$ or $6 \pmod 9$. Since $a^2 \equiv 0, 1, 4$ or $7 \pmod 9$ for any integer a, it follows that $a^2 + b^2 \equiv 0, 1, 2, 4, 5, 7$ or $8 \pmod 9$; hence n is not the sum of two squares.

    (c) If $n = t_k + t_j$ for triangular numbers t_k and t_j, then $n = k(k+1)/2 + j(j+1)/2$. Thus

    $$4n + 1 = (k + j + 1)^2 + (k - j)^2.$$

    (d) For a Fermat number F_n, where $n \geq 1$,

    $$F_n = 2^{2^n} + 1 = (2^{2^{n-1}})^2 + 1^2.$$

    (e) If n is an odd perfect number, then by the corollary to Theorem 11-7, $n = p^k m^2$, where p is prime, $\gcd(p, m) = 1$ and $p \equiv k \equiv 1 \pmod 4$. According to Theorem 13-3, these conditions ensure that n is the sum of two squares.

5. (a) Suppose that $n = 2^m a^2 b$, where $m \geq 0$, a is odd and b has prime divisors of the form $4k + 1$. If m is even, then $n = (2^{m/2})^2 b$ and Theorem 13-3 implies that n is the sum of two squares. If m is odd, then $n = (2^{(m-1)/2})^2(2b)$, so that Theorem 13-3 can again be applied.

(b) Using the identity in Theorem 13-3,

$$
\begin{aligned}
3185 &= 5 \cdot 7^2 \cdot 13 = 7^2(2^2 + 1^2)(2^2 + 3^2) \\
&= 7^2(7^2 + 4^2) = 49^2 + 28^2; \\
39690 &= 2 \cdot 3^4 \cdot 5 \cdot 7^2 = (3^2 \cdot 7)^2(1^2 + 1^2)(2^2 + 1^2) \\
&= (63)^2(3^2 + 1^2) = 189^2 + 63^2; \\
62920 &= 2^3 \cdot 5 \cdot 11^2 \cdot 13 = (2 \cdot 11)^2(3^2 + 1^2)(3^2 + 2^2) \\
&= (22)^2(11^2 + 3^2) = 242^2 + 66^2.
\end{aligned}
$$

7. Suppose that n is not the sum of two squares. By Theorem 13-3, there is a prime $p \equiv 3 \pmod 4$ and an odd integer k satisfying $p^k | n$ and $p^{k+1} \nmid n$. If n could be written as the sum of squares of two rational numbers, say $n = (a/b)^2 + (c/d)^2$, then $n(bd)^2 = (ad)^2 + (bc)^2$. Thus, $n(bd)^2$ can be written as the sum of two squares. But in the prime factorization of $n(bd)^2$, the prime p appears to an odd power, which contradicts the corollary to Theorem 13-3.

9. (a) Let n be a triangular number, say $n = k(k+1)/2$. Clearly, $8n^2 = (2n)^2 + (2n)^2$. Since n is triangular, $8n + 1 = a^2$ by Problem l(b) of Section 1.3. Thus

$$2(8n^2 + 1) = (4n - 1)^2 + (8n + 1) = (4n - 1)^2 + a^2.$$

According to Problem 8, this implies that $8n^2 + 1$ can be written as a sum of two squares. Finally,

$$8n^2 + 2 = [k(k + 1) + 1]^2 + [k(k + 1) - 1]^2,$$

a sum of two squares also.

(b) Given any four consecutive integers, one of them, say n, satisfies $n \equiv 3 \pmod 4$. Write $n = N^2 m$, where m is square-free and $N^2 \equiv 1 \pmod 4$ and $m \equiv 3 \pmod 4$. This means that m contains a prime factor of the form $4k + 3$, whence n cannot be written as a sum of two squares.

11. (a) Let p be an odd prime and $p | (a^2 + b^2)$, where $\gcd(a, b) = 1$. One of a or b is relatively prime to p, say a. Then the congruence $ax \equiv 1 \pmod p$ has a unique solution c. Multiplying $a^2 + b^2 \equiv 0 \pmod p$ by c^2 leads to $(bc)^2 + 1 \equiv 0 \pmod p$, hence $(-1/p) = 1$. By Theorem 9-4, $p \equiv 1 \pmod 4$.

(b) Suppose that $n|(a^2 + b^2)$, where $\gcd(a, b) = 1$. Write $n = N^2m$, with m square-free; in particular, let $m = p_1p_2\cdots p_r$, where $p_1, p_2, \ldots p_r$ are distinct primes Then $p_k|(a^2 + b^2)$, whence by part (a), $p_k \equiv 1 \pmod 4$ for $k = 1, 2, \ldots, r$. Theorem 13-3 now implies that n can be represented as the sum of two squares.

13. (a) Let the positive integer n be the difference of two squares, say $n = a^2 - b^2 = (a+b)(a-b)$. If a and b are both even, or both odd, then $a + b$ and $a - b$ are even. If one of a and b is even with the other odd, then $a+b$ and $a-b$ are both odd. Conversely, assume that $n = cd$ where c and d are both even or both odd. Then $n \equiv 0, 1$ or $3 \pmod 4$. By Theorem 13-4, n can be represented as the difference of two squares.

(b) If n is a positive even integer and $n = a^2 - b^2$, then a and b are both even, or both are odd. If a and b are both even, then $a^2 \equiv b^2 \equiv 0 \pmod 4$, so that $n \equiv 0 \pmod 4$; if a and b are both odd, then $a^2 \equiv b^2 \equiv 1 \pmod 4$, whence $n \equiv 0 \pmod 4$. In either case, $4|n$. Conversely, if $4|n$, Theorem 13-4 implies that n can be represented as the difference of two squares.

15. For any $n > 0$ and $k = 1, 2, \ldots, n$,

$$(2^{2n-k} + 2^{k-1})^2 - (2^{2n-k} - 2^{k-1})^2 = 4 \cdot 2^{2n-k} \cdot 2^{k-1} = 2^{n+1}.$$

Thus $2n + 1$ has n representations as the difference of two squares.

17. If the prime $p \equiv 1$ or $3 \pmod 8$, then by Problem 4 of Section 9.3 the symbol $(-2/p) = 1$; hence, there is some integer a satisfying $a^2 \equiv -2 \pmod p$. Using Thue's Lemma, $ax \equiv y \pmod p$ for suitable integers x, y with $0 < |x| < \sqrt{p}$ and $0 < |y| < \sqrt{p}$. Hence

$$-2x^2 \equiv a^2x^2 \equiv y^2 \pmod p$$

or $2x^2 + y^2 \equiv 0 \pmod p$, so that $2x^2 + y^2 = kp$ for some $k > 0$. But $2x^2 + y^2 < 3p$, hence $k = 1$ or 2. For $k = 1$, it follows that $2x^2 + y^2 = p$; while, for $k = 2$, set $y = 2u$ (since y is even) to get $x^2 + 2u^2 = p$.

13.3   Sums of More Than Two Squares

1. (a) By Problem 5(b) of Section 1.2,

$$1^2 + 2^2 \quad + \quad 3^2 + \cdots + 24^2 = (24 \cdot 25 \cdot 49)/6$$
$$= (2^2 \cdot 5^2 \cdot 7^2) = 70^2.$$

(b)

$$18^2 + 19^2 + 20^2 + \cdots + 28^2$$
$$= (1^2 + 2^2 + \cdots + 28^2) - (1^2 + 2^2 + \cdots + 17^2)$$
$$= (28 \cdot 29 \cdot 57)/6 - (17 \cdot 18 \cdot 35)/6$$
$$= 7714 - 1785 = 5929 = 77^2.$$

(c)

$$2^2 + 5^2 + 8^2 + \cdots + 23^2$$
$$= 2^2 + (2+3)^2 + (2 + 2 \cdot 3)^2 + \cdots + (2 + 8 \cdot 3)^2$$
$$= 9 \cdot 2^2 + 4(1 + 2 + \cdots + 8)3 + (1^2 + 2^2 + \cdots + 8^2)3^2$$
$$= 3^2(4 + 48 + 204) = 3^2 \cdot 16^2 = 48^2.$$

(d)

$$6^2 + 12^2 + 18^2 + \cdots + 48^2$$
$$= 6^2(1^2 + 2^2 + \cdots + 8^2)$$
$$= (6^2 \cdot 8 \cdot 9 \cdot 17)/6$$
$$= 7344 = 136 \cdot 54$$
$$= \left(\frac{136 + 54}{2}\right)^2 - \left(\frac{136 - 54}{2}\right)^2 = 95^2 - 41^2.$$

3. Assume to the contrary that $q_i \neq 3$ for $i = 1, 2, 3$. Then each q_i is either 2 or is an integer of the form $3k \pm 1$. By considering the various cases, we conclude that

$$p = q_1^2 + q_2^2 + q_3^2 \equiv 0 \pmod{3}$$

and so 3 divides the prime p. But this is impossible, as $p > 3$. Hence, $q_i = 3$ for some value of i.

5. If n is not a sum of three squares, then $n = 4^k(8m + 7)$ for some nonnegative k and m. Then $2n = 2 \cdot 4^k(8m + 7)$ is not of the form $4^j(8r + 7)$. Indeed, if $2 \cdot 4^k(8m + 7) = 4^j(8r + 7)$, with say $k \geq j$, then

$$6 \cdot 4^{k-j} \equiv 2 \cdot 4^{k-j}(8m + 7) \equiv 8r + 7 \equiv 7 \pmod{8}.$$

Thus, $2n$ can be written as a sum of three squares.

9. (a) For an integer a, $a^3 \equiv 0, 1$ or 8 (mod 9). Thus
$$a^3 + b^3 + c^3 \equiv 0, 1, 2, 3, 6, 7 \text{ or } 8 \pmod{9}.$$
This means that if $n \equiv 4$ or 5 (mod 9), then it cannot be a sum of three or fewer cubes.

   (b) Let the prime $p = a^3 + b^3$ for some $a \geq 0$, $b \geq 0$. Since
$$a^3 + b^3 = (a + b)(a^2 - ab + b^2) = (a + b)((a - b)^2 + ab)$$
either $a + b = 1$ or $(a - b)^2 + ab = 1$. If $a + b = 1$, then either $a = 0$ or $b = 0$, so that either $p = a^3$ or $p = b^3$, which is impossible. Thus $(a - b)^2 + ab = 1$, which implies that $a = b = 1$. That is, $p = 1^3 + 1^3 = 2$.

   (c) Let the prime $p = a^3 - b^3 = (a - b)(a^2 + ab + b^2)$ for some $a \geq 0$, $b \geq 0$. Then $a - b = 1$ and $a^2 + ab + b^2 = p$. Since $a = b + 1$,
$$p = (b + 1)^2 + (b + 1)b + b^2 = 3b(b + 1) + 1.$$
Conversely, if p is of the form $p = 3k(k + 1) + 1$, then
$$p = (k + 1)^3 - k^3,$$
the difference between the cubes of two positive integers.

11. Since $8n + 3$ is not of the form $4^k(8m + 7)$, it can be written as the sum of three squares. Now $8n + 3 = x^2 + y^2 + z^2$ implies that x, y, z are all odd or exactly two of them are even. If two are even, say x and y, then $8n + 3 = 4x_1^2 + 4y_1^2 + z^2$: hence $z^2 \equiv 3$ (mod 4), which cannot occur. It follows that $8n + 3$ can be represented as
$$
\begin{aligned}
8n + 3 &= (2a + 1)^2 + (2b + 1)^2 + (2c + 1)^2 \\
&= 4a(a + 1) + 4b(b + 1) + 4c(c + 1) + 3
\end{aligned}
$$
or
$$n = \frac{a(a + 1)}{2} + \frac{b(b + 1)}{2} + \frac{c(c + 1)}{2},$$
a sum of three triangular numbers.

15. Let $n^3 \equiv n$ (mod 6), so that $n - n^3 = 6k$ for some k. Then
$$
\begin{aligned}
n &= n^3 + 6k \\
&= n^3 + (k + 1)^3 + (k - 1)^3 - k^3 - k^3 \\
&= n^3 + (k + 1)^3 + (k - 1)^3 + (-k)^3 + (-k)^3,
\end{aligned}
$$
a sum of five cubes, allowing negative cubes.

17. In the identity $(27k^6)^2 - 1 = (9k^4 - 3k)^3 + (9k^3 - 1)^3$, let k be an odd integer. Then $27k^6$ will be odd, call it $2n + 1$. Also $9k^4$ and $3k$ are both odd, so that $9k^4 - 3k$ is even, say $2a$: similarly, $9k^3 - 1$ is even, say $2b$. Thus the identity becomes $(2n + 1)^2 - 1 = (2a)^3 + (2b)^3$. For the triangular number t_n,

$$t_n = \frac{n(n + 1)}{2} = \frac{(2n + 1)^2 - 1}{8} = a^3 + b^3.$$

A similar argument using $(27k^6)^2 - 1 = (9k^4 + 3k)^3 - (9k^3 + 1)^3$ shows that $t + n$ is also the difference of two cubes. Taking $k = 1$ leads to $n = 13$:

$$t_{13} = 91 = 3^3 + 4^3 = 6^3 - 5^3.$$

Chapter 14

Fibonacci Numbers

14.2   The Fibonacci Sequence

1. Note that 7 divides $u_8 = 21$, 11 divides $u_{10} = 55$, 13 divides $u_{14} = 377$ and 17 divides $u_{18} = 2584$.

3. The identity $u_{n+1}^2 - u_{n-1}^2 = u_n^2 + 2u_n u_{n-1}$ implies that if $2|u_n$, then $4|(u_{n+1}^2 - u_{n-1}^2)$. Similarly, from $u_{n+1}^3 - u_{n-1}^3 = u_n^3 + 3u_n^2 u_{n-1} + 3u_n u_{n-1}^2$, it can be concluded that if $3|u_n$, then $9|(u_{n+1}^3 - u_{n-1}^3)$.

5. If $k \geq 2$, $u_k^2 = u_k u_{k+1} - u_k u_{k-1}$. Thus

$$
\begin{aligned}
u_1^2 \; + \; & u_2^2 + u_3^2 + \cdots + u_n^2 \\
= \; & u_1^2 + (u_2 u_3 - u_2 u_1) + (u_3 u_4 - u_3 u_2) + \cdots + (u_n u_{n+1} - u_n u_{n-1}) \\
= \; & u_1^2 - u_2 u_1 + u_n u_{n+1} \\
= \; & u_n u_{n+1}.
\end{aligned}
$$

7. Since $\gcd(9, 12) = 3$, it follows that $\gcd(u_9, u_{12}) = u_3 = 2$. Similarly, $\gcd(15, 20) = 5$ yields $\gcd(u_{15}, u_{20}) = u_5 = 5$, while $\gcd(24, 36) = 12$ implies $\gcd(u_2, u_{36}) = u_{12} = 144$.

9. (a) $2|u_n$, or rather $u_3|u_n$, if and only if $3|n$.

   (b) $3|u_n$, or rather $u_4|u_n$, if and only if $4|n$.

   (c) If $6|n$, then $u_6|u_n$ or $8|u_n$, whence $4|u_n$.

13. The asserted congruence is clearly true for $n = 1$, so assume it holds for $n \leq k$; that is, $2^{k-1} u_k \equiv k \pmod 5$, and also $2^{k-2} u_{k-1} \equiv k - 1$

(mod 5). Then

$$
\begin{aligned}
2^{k-2}u_{k+1} &= 2^k(u_k + u_{k-1}) \\
&= 2(2^{k-1}u_k) + 2^2(2^{k-2}u_{k-1}) \\
&\equiv 2k + 4(k-1) \equiv 6k - 4 \equiv k+1 \pmod 5
\end{aligned}
$$

whence the assertion holds for $k+1$.

15. Suppose that there is some integer n for which

$$
u_1 + u_2 + \cdots + u_{3n} = 16!,
$$

or equivalently by identity (20) that $u_{3n+2} - 1 = 16!$. Using Wilson's Theorem, it follows that $u_{3n+2} - 1 \equiv -1 \pmod{17}$, whence $u_{3n+2} \equiv 0 \pmod{17}$. This gives $17 | u_{3n+2}$, or $u_9 | u_{3n+2}$. ¿From the corollary to Theorem 14-1, $9 | 3n + 2$, which is impossible since $3n + 2 \equiv 2, 5 \text{ or } 8 \pmod 9$.

17. For $n \geq 1$, the n consecutive Fibonacci numbers

$$
u_{(n+2)!+3}, \; u_{(n+2)!+4}, \dots, u_{(n+2)!+(n+2)}
$$

are divisible by $u_3, u_4, \dots, u_{n+2}$ respectively.

19. The asserted congruence clearly holds for $n = 1$, so assume that it holds for all $n \leq k$; then

$$
u_{2k} \equiv k(-1)^{k+1} \pmod 5 \quad \text{and} \quad u_{2k-2} \equiv (k-1)(-1)^k \pmod 5.
$$

For $n = k+1$,

$$
\begin{aligned}
u_{2(k+1)} &= 3u_{2k} - u_{2k-2} \\
&\equiv 3k(-1)^{k+1} - (k-1)(-1)^k \\
&\equiv (-4k+1)(-1)^{k+2} \equiv (k+1)(-1)^{k+2} \pmod 5.
\end{aligned}
$$

Thus the assertion is also true for $n = k+1$.

14.3   Certain Identities Involving Fibonacci Numbers

1. (a) The asserted identity is clearly true for $n = 1$; that is,

$$
1 = u_1 = 2u_3 - u_5 + 2 = 2 \cdot 2 - 5 + 2.
$$

Assume that it holds for $k > 1$, so that

$$u_1 + 2u_2 + \cdots + ku_k = (k+1)u_{k+2} - u_{k+4} + 2.$$

For $k + 1$,

$$
\begin{aligned}
u_1 &+ 2u_2 + \cdots + ku_k + (k+1)u_{k+1} \\
&= (k+1)u_{k+2} - u_{k+4} + 2 + (k+1)u_{k+1} \\
&= (k+1)(u_{k+2} + u_{k+1}) - u_{k+4} + 2 \\
&= (k+1)u_{k+3} + u_{k+3} - (u_{k+4} + u_{k+3}) + 2 \\
&= (k+2)u_{k+3} - u_{k+5} + 2
\end{aligned}
$$

and so the identity holds in this case also.

(b) Since $1 = u_2 = 1 \cdot u_3 - u_2 = 2 - 1$, the asserted identity holds for $n = 1$. Assuming it is true for k, that is,

$$u_2 + 2u_4 + 3u_6 + \cdots + ku_{2k} = ku_{2k+1} - u_{2k},$$

consider the case of $k + 1$. Then

$$
\begin{aligned}
u_2 + 2u_4 &+ \cdots + ku_{2k} + (k+1)u_{2(k+1)} \\
&= ku_{2k+1} - u_{2k} + (k+1)u_{2k+2} \\
&= k(u_{2k+1} + u_{2k+2}) - u_{2k} + u_{2k+2} \\
&= ku_{2k+3} - u_{2k} + u_{2k+2} \\
&= (k+1)u_{2k+3} - u_{2k} + u_{2k+2} - u_{2k+3} \\
&= (k+1)u_{2k+3} - u_{2k} + u_{2k+2} - (u_{2k+2} + u_{2k1}) \\
&= (k+1)u_{2k+3} - u_{2k+2}
\end{aligned}
$$

and the result holds for $k + 1$.

3. From formula (1), for $n \geq 2$,

$$u_{2n-1} = u_{n+(n-1)} = u_{n-1}u_{n-1} + u_n u_n = u_{n-1}^2 + u_n^2,$$

and

$$
\begin{aligned}
u_{2n} &= u_{2n+1} - u_{2n-1} \\
&= (u_{n+1}^2 + u_n^2) - (u_n^2 + u_{n-1}^2) = u_{n+1}^2 - u_{n-1}^2.
\end{aligned}
$$

5. Using identity (3),

$$
\begin{aligned}
u_n u_{n-1} &= (u_{n+1} - u_{n-1})u_{n-1} \\
&= u_{n+1}u_{n-1} - u_{n-1}^2 \\
&= u_{n+1}u_{n-1} + (-1)^{n-1} - u_{n-1}^2 + (-1)^n \\
&= u_n^2 - u_{n-1}^2 + (-1)^n.
\end{aligned}
$$

If $d = \gcd(u_n, u_{n-1})$, then

$$
d \mid (u_n u_{n-1} - u_n^2 + u_{n-1}^2)
$$

or $d \mid (-1)^n$. Since $d > 0$, it follows that $d = 1$.

9. Note that

$$
\begin{aligned}
u_{n+2}^2 - u_{n+1}^2 &= (u_{n+3} + u_{n+1})^2 - u_{m+1}^2 \\
&= u_{n+3}(u_{n+3} - 2u_{n+1}) \\
&= u_{n+3}(u_{n+2} + u_{n+1} - 2u_{n+1}) \\
&= u_{n+3}(u_{n+2} - u_{n+1}) = u_{n+3}u_n.
\end{aligned}
$$

From this, it follows upon using Problem 3 that

$$
\begin{aligned}
u_{2n+3}^2 &= (u_{n+2}^2 + u_{n+1}^2)^2 \\
&= (u_{n+2}^2 - u_{n+1}^2)^2 + 4u_{n+1}^2 u_{n+2}^2 \\
&= (u_{n+3}u_n)^2 + (2u_{n+1}u_{n+2})^2.
\end{aligned}
$$

In particular,

$$
\begin{array}{lll}
(u_4 u_1)^2 + (2u_2 u_3)^2 = u_5^2 & \text{or} & 3^2 + 4^2 = 5^2; \\
(u_5 u_2)^2 + (2u_3 u_4)^2 = u_7^2 & \text{or} & 5^2 + 12^2 = 13^2; \\
(u_7 u_4)^2 + (2u_5 u_6)^2 = u_{11}^2 & \text{or} & 39^2 + 80^2 = 89^2.
\end{array}
$$

11. For $n \geq 1$,

$$
\begin{aligned}
u_{2n+2}u_{2n-1} &- u_{2n}u_{2n+1} \\
&= (\alpha^{2n+2} - \beta^{2n+2})(\alpha^{2n-1} - \beta^{2n-1}) \\
&\quad - (\alpha^{2n} - \beta^{2n})(\alpha^{2n+1} - \beta^{2n+l})(\alpha - \beta)^2 \\
&= \frac{-(\alpha\beta)^{2n-1}(\alpha^2 + \beta^3) + (\alpha\beta)^{2n}(\alpha + \beta)}{5} \\
&= \frac{\alpha^3 + \beta^3 + 1}{5} = \frac{2(\alpha + \beta) + 2 + 1}{5} = 5,
\end{aligned}
$$

since $\alpha^3 = 2\alpha + 1$ and $\beta^3 = 2\beta + 1$.

13. (a) Let $p = 4k+3$ be prime and assume to the contrary that $p|u_{2n-1}$ for some $n \geq 1$. Then, by Problem 3,

$$u_n^2 + u_{n-1}^2 = u_{2n-1} \equiv 0 \pmod{p}.$$

From Problem 12 of Section 5.4, it follows that

$$u_n \equiv u_{n-1} \equiv 0 \pmod{p},$$

whence $p|u_n$ and $p|u_{n-1}$. But this contradicts the fact that successive Fibonacci numbers are relatively prime.

(b) Consider the sequence $\{u_p\}$, where $p > 5$ is prime. By part (a), each term of the sequence is divisible by a prime of the form $4k + 1$. But from Theorem 14-3, the terms of the sequence are relatively prime to each other, hence have distinct prime divisors. Thus, there are infinitely many primes of the form $4k + 1$.

15. Given any twenty consecutive Fibonacci numbers $u_{n+1}, u_{n+2}, \ldots, u_{n+20}$, identity (1) gives

$$
\begin{aligned}
u_{n+1} \ + \ & u_{n+2} + \cdots + u_{n+20} \\
= \ & (u_{n-1}u_1 + u_n u_2) + (u_{n-1}u_2 + u_n u_3) + \cdots + (u_{n-1}u_{20} + u_n u_{21}) \\
= \ & u_{n-1}(u_1 + u_2 + \cdots + u_{20}) + u_n(u_2 + u_3 + \cdots + u_{21})
\end{aligned}
$$

where both expressions in parentheses are divisible by u_{10}. For, using Problem 3 and identity (3),

$$
\begin{aligned}
u_1 + u_2 + \cdots + u_{20} \ = \ & u_{22} - 1 = u_{20} + (u_{21} - 1) \\
= \ & u_{20} + (u_{10}^2 + u_9^2 - 1) \\
= \ & u_{20} + u_{10}^2 + u_{10}u_8
\end{aligned}
$$

where $u_{10}|u_{20}$. Also,

$$
\begin{aligned}
u_2 + u_3 \ + \ & \cdots + u_{20} + u_{21} \\
= \ & (u_1 + u_2 + \cdots + u_{20}) + (u_{21} - 1) \\
= \ & (u_1 + u_2 + \cdots + u_{20}) + u_{10}^2 + u_{10}u_8.
\end{aligned}
$$

17. (a) The identity holds for $n = 1$, since $L_1 = 1 = 4 - 3 = L_3 - 3$. Assume that it also holds for $n = k$, so that

$$L_1 + L_2 + L_3 + \cdots + L_k = L_{k+2} - 3.$$

Then

$$L_1 + L_2 \; + \; L_3 + \cdots + L_k + L_{k+1}$$
$$= (L_{k+2} - 3) + L_{k+1}$$
$$= (L_{k+2} + L_{k+1}) - 3 = L_{k+3} - 3;$$

hence the identity holds for $n = k + 1$ also, and thus for all n by induction.

(b) When $n = 1$, then $L_1 = 1 = 3 - 2 = L_2 - 2$, so the assertion is correct in this case. Suppose that for some $k_{,,}$

$$L_1 + L_3 + L_5 + \cdots + L_{2k-1} = L_{2k} - 2.$$

Then

$$L_1 + L_3 \; + \; L_5 + \cdots + L_{2k-1} + L_{2k+1} = (L_{2k} - 2) + L_{2k+1}$$
$$= (L_{2k+1} + L_{2k}) - 2 = L_{2(k+1)} - 2,$$

which completes the induction step.

(c) When $n = 1$, the assertion reduces to $L_2 = 3 = 4 - 1 = L_3 - 1$, which is clearly true. For some k, assume that

$$L_2 + L_4 + L_6 + \cdots + L_{2k} = L_{2k+1} - 1.$$

Then

$$L_2 + L_4 \; + \; L_6 + \cdots + L_{2k} + L_{2k+2} = (L_{2k+1} - 1) + L_{2k+2}$$
$$= (L_{2k+2} + L_{2k+1}) - 1 = L_{2k+3} - 1,$$

showing that the assertion holds for $k + 1$ whenever it holds for k.

(d) Since $L_3^3 = 4^2 = 7 \cdot 3 + 5(-1)^3 = L_4 L_2 + 5(-1)^3$, the identity holds for $n = 3$. Assume it also holds for $n = k$, where $k \geq 3$, so that $L_k^2 = L_{k+1} L_{k-1} + 5(-1)^k$. Then

$$\begin{aligned} L_{k+1}^2 &= (L_k + L_{k-1})(L_{k+2} - L_k) \\ &= L_k L_{k+2} + L_{k-1} L_{k+2} - L_{k-1} L_k - L_k^2 \\ &= L_k L_{k+2} + L_{k-1}(L_{k+2} - L_k - L_{k+1})5(-1)^{k+1} \\ &= L_k L_{k+2} + 5(-1)^{k+1}. \end{aligned}$$

Thus the assertion holds for $n = k + 1$.

(e) When $n = 1$, $L_1^2 = 1 = 1 \cdot 3 - 2 = L_1 L_2 - 2$, whence the identity holds in this case. For some k, assume that

$$L_1^2 + L_2^2 + L_3^2 + \cdots + L_k^2 = L_k L_{k+1} - 2.$$

Then

$$
\begin{aligned}
L_1^2 \;+\; & L_2^2 + L_3^2 + \cdots + L_k^2 + L_{k+1}^2 \\
= \;& L_k L_{k+1} - 2 + L_{k+1}^2 \\
= \;& L_k L_{k+1} - 2 + L_{k+1}(L_{k+2} - L_k) \\
= \;& L_{k+1} L_{k+2} - 2,
\end{aligned}
$$

completing the induction step.

(f) For $n \geq 2$,

$$
\begin{aligned}
L_{n+1}^2 - L_n^2 &= (L_n + L_{n-1})^2 - L_n^2 \\
&= L_{n-1}^2 + 2 L_n L_{n-1} \\
&= (L_n + L_{n+1}) L_{n-1} = L_{n+2} L_{n-1}
\end{aligned}
$$

19. Using part (a) of Problem 18 and the Binet formula for Fibonacci numbers,

$$
\begin{aligned}
L_n &= u_{n+1} + u_{n-1} \\
&= \frac{\alpha^{n+1} - \beta^{n+1}}{\alpha - \beta} + \frac{\alpha^{n-1} - \beta^{n-1}}{\alpha - \beta} \\
&= \frac{\alpha^{n-1}(\alpha^2 + 1) + \beta^{n-1}(\beta^2 + 1)}{\alpha - \beta} \\
&= \frac{\alpha^{n-1} \cdot \alpha(\alpha - \beta) + \beta^{n-1} \cdot \beta(\alpha - \beta)}{\alpha - \beta} \\
&= \alpha^n + \beta^n.
\end{aligned}
$$

21. (a) For $n \geq 1$,

$$
\begin{aligned}
L_n^2 - 5 u_n^2 &= (\alpha^n + \beta^n)^2 - (\alpha^n - \beta^n)^2 \\
&= (\alpha^{2n} + 2(-1)^n + \beta^{2n}) + (\alpha^{2n} - 2(-1)^n + \beta^{2n}) \\
&= 4(-1)^n.
\end{aligned}
$$

(b) For $n \geq 1$,

$$
\begin{aligned}
L_{2n+1} - 5 u_n u_{n+1} &= (\alpha^{2n+1} + \beta^{2n+1}) - (\alpha^n - \beta^n)(\alpha^{n+1} + \beta^{n+1}) \\
&= \alpha(\alpha\beta)^n + \beta(\alpha\beta)^n \\
&= \alpha(-1)^n + \beta(-1)^n = (\alpha + \beta)(-1)^n = (-1)^n
\end{aligned}
$$

(c) For $n \geq 1$,

$$
\begin{aligned}
L_n^2 - u_n^2 &= (\alpha^n + \beta^n)^2 - \left(\frac{\alpha^n - \beta^n}{\sqrt{5}}\right)^2 \\
&= (4/5)(\alpha^{2n} + 3(-1)^n + \beta^{2n}) \\
&= (4/5)(\alpha^{2n} + (\alpha^2 + \beta^2)(-1)^{n-1} + \beta^{2n}) \\
&= 4\left(\frac{\alpha^{n-1} - \beta^{n-1}}{\sqrt{5}}\right)\left(\frac{\alpha^{n+1} - \beta^{n+1}}{\sqrt{5}}\right) \\
&= 4u_{n-1}u_{n+1}
\end{aligned}
$$

(d) For $n \geq 1$,

$$
\begin{aligned}
L_m L_n + 5u_m u_n &= (\alpha^m + \beta^m)(\alpha^n + \beta^n) + (\alpha^m - \beta^m)(\alpha^n - \beta^n) \\
&= 2(\alpha^{m+n} + \beta^{m+n}) = 2L_{m+n}.
\end{aligned}
$$

23. The asserted equality holds for $n = 1$ and 2, since

$$
u_1 = 1 = \begin{pmatrix} 0 \\ 0 \end{pmatrix}, \quad u_2 = 1 = \begin{pmatrix} 1 \\ 0 \end{pmatrix},
$$

Assume the equality holds for all u_n where $n < k$, so that

$$
\begin{aligned}
u_k &= \begin{pmatrix} k-1 \\ 0 \end{pmatrix} + \begin{pmatrix} k-2 \\ 1 \end{pmatrix} + \begin{pmatrix} k-3 \\ 2 \end{pmatrix} \\
&\quad + \cdots + \begin{pmatrix} k-j \\ j-1 \end{pmatrix} + \begin{pmatrix} k-j-1 \\ j \end{pmatrix}, \\
&\text{where} \quad j = \left[\frac{k-1}{2}\right],
\end{aligned}
$$

and

$$
\begin{aligned}
u_{k-1} &= \begin{pmatrix} k-2 \\ 0 \end{pmatrix} + \begin{pmatrix} k-3 \\ 1 \end{pmatrix} + \begin{pmatrix} k-4 \\ 2 \end{pmatrix} \\
&\quad + \cdots + \begin{pmatrix} k-i-1 \\ i-1 \end{pmatrix} + \begin{pmatrix} k-i-2 \\ i \end{pmatrix}, \\
&\text{where} \quad i = \left[\frac{k-2}{2}\right].
\end{aligned}
$$

Now consider the case where $n = k + 1$. If k is taken to be even, say $k = 2m$, then $i = [(k-2)/2] = m - 1 = [(k-1)/2] = j$ and

$r = [k/2] = m = j + 1$. Thus,

$$
\begin{aligned}
u &= u_k + u_{k+1} \\
&= \binom{k-1}{0} + \left[\binom{k-2}{1} + \binom{k-2}{0}\right] + \cdots + \\
&\quad \left[\binom{k-j-1}{j} + \binom{k-j-1}{j-1}\right] + \binom{k-j-2}{j} \\
&= \binom{k-1}{0} + \binom{k-1}{1} + \cdots + \binom{k-j}{j} + \binom{k-j-2}{j} \\
&= \binom{k}{0} + \binom{k-1}{1} + \cdots + \binom{k-r+1}{r-1} + \binom{k-r-1}{r-1} \\
&= \binom{k}{0} + \binom{k-1}{1} + \cdots + \binom{k-r+1}{r-1} + \binom{k-r}{r},
\end{aligned}
$$

since

$$
\binom{k-r+1}{r-1} = 1 = \binom{k-r}{r}.
$$

Thus the equality holds for $n = k + 1$ also. If k is an odd integer, a similar argument establishes the equality.

25. The identity

$$
u_n + u_{n+1} + \cdots + u_{n+k-1} = u_{n-1}(u_{k+1} - 1) + u_n(u_{k+2} - 1), n \geq 2
$$

is established by induction on $k \geq 1$. It clearly holds for $k = 1$, since

$$
u_n = u_{n-1}(u_2 - 1) + u_n(u_3 - 1) = u_{n-1} \cdot 0 + u_n \cdot 1.
$$

Assume that the identity holds for some k. Then using identity (1),

$$
\begin{aligned}
u_n + u_{n+1} &+ \cdots + u_{n+k-1} + u_{n+k} \\
&= u_{n-1}(u_{k+1} - 1) + u_n(u_{k+2} - 1) + u_{n+k} \\
&= u_{n-1}(u_{k+1} - 1) + u_n(u_{k+2} - 1) + (u_{n-1}u_k + u_n u_{k+1}) \\
&= u_{n-1}(u_{k+1} + u_k - 1) + u_n(u_{k+2} + u_{k+1} - 1) \\
&= u_{n-1}(u_{k+2} - 1) + u_n(u_{k+3} - 1),
\end{aligned}
$$

whence the asserted equality also holds for $k + 1$. For $n \geq 2$ and $k = 24$, it follows that

$$
\begin{aligned}
u_n + u_{n+1} &+ \cdots + u_{n+23} = u_{n-1}(u_{25} - 1) + u_n(u_{26} - 1) \\
&= u_{n-1}(24 \cdot 3126) + u_n(24 \cdot 5058),
\end{aligned}
$$

and so the sum is divisible by 24. The sum of the first 24 Fibonacci numbers is also divisible by 24, since

$$u_1 + u_2 + \cdots + u_{24} = u_{25} - 1 = 24 \cdot 5058.$$

27. For $n \geq 1$,

$$
\begin{aligned}
\frac{u_{n+1}}{u_n} &= \left(\frac{\alpha^{n+1}}{\sqrt{5}} + \delta_{n+1} \right) \bigg/ \left(\frac{\alpha^n}{\sqrt{5}} + \delta_n \right) \\
&= \left(\alpha + \frac{\sqrt{5}\delta_{n+1}}{\alpha^n} \right) \bigg/ \left(1 + \frac{\sqrt{5}\delta_n}{\alpha^n} \right).
\end{aligned}
$$

Taking limits, it follows that

$$\lim_{n \to \infty} \frac{u_{n+1}}{u_n} = \frac{\alpha + 0}{1 + 0} = \alpha.$$

Chapter 15

Continued Fractions

15.2 Finite Continued Fractions

1. (a) $-19/51 = -1 + 32/51$. The Euclidean Algorithm yields $51 = 1 \cdot 32 + 19$, $32 = 1 \cdot 19 + 13$, $19 = 1 \cdot 13 + 6$, $13 = 2 \cdot 6 + 1$, $6 = 6 \cdot 1 + 0$. Thus, $-19/51 = [-1; 1, 1, 1, 2, 6]$.

   (b) $187/57 = [3; 3, 1, 1, 3, 2]$.

   (c) $71/55 = [1; 3, 2, 3, 2]$.

   (d) $119/303 = [0; 2, 1, 1, 3, 5, 3]$.

3. Let $r = [a_0; a_1, a_2, \ldots, a_n]$, where $r > 1$. Then

$$r = a_0 + \cfrac{1}{a_1 + \cfrac{1}{a_2 + \cfrac{\ddots}{+\frac{1}{a_n}}}}, \quad \text{so that} \quad \frac{1}{r} = \cfrac{1}{a_0 + \cfrac{1}{a_1 + \cfrac{1}{a_2 + \cfrac{\ddots}{+\frac{1}{a_n}}}}}.$$

   Hence, $1/r = [0; a_0, a_1, a_2, \ldots, a_n]$.

5. (a) For $[1; 2, 3, 3, 2, 1]$, the convergents are $C_0 = 1$, $C_1 = 3/2$, $C_2 = 10/7$, $C_3 = 33/23$, $C_4 = 76/53$, $C_5 = 109/76$.

   (b) For $[-3; 1, 1, 1, 1, 3]$, the convergents are $C_0 = -3$, $C_1 = -2$, $C_2 = -5/2$, $C_3 = -7/3$, $C_4 - -12/5$, $C_5 = -43/18$.

   (c) For $[0; 2, 4, 1, 8, 2]$, the convergents are $C_0 = 0$, $C_1 = 1/2$, $C_2 = 4/9$, $C_3 = 5/11$, $C_4 = 44/97$, $C_5 = 93/205$.

117

7. (a) For $[1; 2, 2, 2, 2, 2, 2, 2, 2]$,
        the $p_k (0 \leq k \leq 8)$ are $1, 3, 7, 17, 41, 99, 239, 577, 1393$; and
        the $q_k (0 \leq k \leq 8)$ are $1, 2, 5, 12, 29, 70, 169, 408, 985$.
        Hence the convergents are
        $1, 3/2, 7/5, 17/12, 41/29, 99/70, 239/169, 577/408, 1393/985$.

    (b) For $[1; 1, 2, 1, 2, 1, 2, 1, 2]$,
        the p_k $(0 \leq k \leq 8)$ are $1, 2, 5, 7, 19, 26, 71, 97, 265$; and
        the q_k $(0 \leq k \leq 8)$ are $1, 1, 3, 4, 11, 15, 41, 56, 153$.
        Thus the convergents are
        $1, 2, 5/3, 7/4, 19/11, 26/15, 71/41, 97/56, 265/153$.

    (c) For $[2; 4, 4, 4, 4, 4, 4, 4, 4]$,
        the $p_k (0 \leq k \leq 8)$ are $2, 9, 38, 161, 682, 2889, 12238, 51841, 219602$;
        the $q_k (0 \leq k \leq 8)$ are $1, 4, 17, 72, 305, 1292, 5473, 23184, 98209$.
        The corresponding convergents are
        $2, 9/4, 38/17, 161/72, 682/304, 2889/1292, 12238/5473,$
        $51841/23184, 219602/98209$.

    (d) For $[2; 2, 4, 2, 4, 2, 4, 2, 4]$,
        the $p_k (0 \leq k \leq 8)$ are $2, 5, 22, 49, 218, 485, 2158, 4801, 21362$;
        the $q_k (0 \leq k \leq 8)$ are $1, 2, 9, 20, 89, 198, 881, 1960, 8721$.
        These produce the convergents
        $2, 5/2, 22/9, 49/20, 218/89, 485/198, 2158/881, 4801/1960,$
        $21362/8721$.

    (e) For $[2; 1, 1, 1, 4, 1, 1, 1, 4]$,
        the $p_k (0 \leq k \leq 8)$ are $2, 3, 5, 8, 37, 45, 82, 127, 590$; and
        the $q_k (0 \leq k \leq 8)$ are $1, 1, 2, 3, 14, 17, 31, 48, 223$.
        Thus the convergents are
        $2, 3, 5/2, 8/3, 37/14, 45/17, 82/31, 127/48, 590/223$.

9. $3.1416 = 3 + (1416/10000) = 3 + (177/1250) = [3; 7, 16, 11]$, since
    $1250 = 177 \cdot 7 + 11, \quad 177 = 11 \cdot 16 + 1$.
    Similarly, $3.14159 = 3 + (14159/100000) = [3; 7, 15, 1, 25, 1, 7, 4]$.

11. (a) For the equation $19x + 51y = 1$, $19/51 = [0; 2, 1, 2, 6]$. A partic-
        ular solution of the equation is $x_0 = -q_3 = -8$, $y_0 = p_3 = 3$, so
        the general solution is $x = -8 + 51t$, $y = 3 - 19t$ for $t = 0, \pm 1, \ldots$.

    (b) Given $364x + 227y = 1$, we have $364/227 = [1; 1, 1, 1, 1, 10, 1, 3]$.
        Since one solution of the equation is $x_0 = q_6 = 58$, $y_0 = -p_6 = -93$, its general solution will be $x = 58 + 227t$, $y = -93 - 364t$,
        where $t = 0, \pm 1, \ldots$.

(c) For $18x + 5y = 24$, $18/5 = [3; 1, 1, 1]$. A particular solution of the equation is $x_0 = 24q_2 = 24 \cdot 2 = 48$, $y_0 = 24(-p_2) = 24(-7) = -168$. Thus the general solution is $x = 48 + 5t$, $y = -168 - 18t$ for $t = 0, \pm 1, \ldots$.

(d) Consider the equation $158x - 57y = 1$. Here, $158/57 = [2; 1, 3, 2, 1, 1, 1]$. Since $x_0 = -q_5 = -22$, $y_0 = -p_5 = -61$ is one solution of the equation, its general solution is $x = -22 - 57t$, $y = -61 - 158t$, where $t = 0, \pm 1, \ldots$.

15.3 Infinite Continued Fractions

1. (a) Let $x = [\overline{2; 3}]$. Then $x = [2; 3, x]$, so that

$$x = 2 + \frac{1}{3 + (1/x)} = 2 + \frac{x}{3x + 1} = \frac{7x + 2}{3x + 1}$$

Thus $3x^2 - 6x - 2 = 0$ and, since x is positive, $x = \frac{3 + \sqrt{15}}{2}$.

(b) Let $x = [0; \overline{1, 2, 3}]$. Then $x = [0; y]$, where $y = [\overline{1; 2, 3,}] = [1; 2, 3, y]$. Therefore,

$$y = 1 + \frac{1}{2 + \frac{1}{3 + (1/y)}} = \frac{10y + 3}{7y + 2},$$

and so $7y^2 - 8y - 3 = 0$. Because $y > 0$, we have $y = \frac{4 + \sqrt{37}}{7}$, which yields $x = 1/y = \frac{\sqrt{37} - 4}{3}$.

(c) Let $x = [2; \overline{1, 2, 1}]$. Then $x = [2; y]$, with $y = [\overline{1; 2, 1}] = [1; 2, 1, y]$. This implies that

$$y = 1 + \frac{1}{2 + \frac{1}{1 + (1/y)}} = \frac{4y + 3}{3y + 2}.$$

Then $3y^2 - 2y - 3 = 0$ and y is positive, giving $y = \frac{1 + \sqrt{10}}{3}$. It follows that

$$x = 2 + 1/y = 2 + \frac{3}{1 + \sqrt{10}} = \frac{5 + \sqrt{10}}{3}.$$

(d) Let $x = [1; 2, \overline{3, 1}] = [1; 2, y]$, where $y = [\overline{3; 1}] = [3; 1, y]$. Then

$$y = 3 + \frac{1}{1 + (1/y)} = \frac{4y + 3}{y + 1}$$

or $y^2 - 3y - 3 = 0$. Then $y = \frac{3+\sqrt{21}}{2}$, which means that

$$x = 1 + \cfrac{1}{2 + (1/y)} = \frac{19 - \sqrt{21}}{10}.$$

(e) Let $x = [1; 2, 1, 2, \overline{12}] = [1; 2, 1, 2, y]$, where $y = [\overline{12}; 0] = [12; y]$. Then

$$y = 12 + (1/y) = \frac{12y + 1}{y},$$

or $y^2 - 12y - 1 = 0$. This given $y = 6 + \sqrt{37}$ and $x = \frac{314+\sqrt{37}}{233}$.

3. Let $x = [1; 2, \overline{1}] = [1; 2, y]$, where $y = [1; \overline{1}]$. From Problem 2, $y = (1 + \sqrt{5})/2$, so that

$$x = 1 + \cfrac{1}{2 + (1/y)} = \frac{5 - \sqrt{5}}{2}.$$

Also

$$
\begin{aligned}
z = [1; 2, 3, \overline{1}] &= [1; 2, 3, y] = 1 + \cfrac{1}{2 + \cfrac{1}{3 + (1/y)}} \\
&= \frac{87 + \sqrt{5}}{62}.
\end{aligned}
$$

5. (a) For $\sqrt{n^2 + 1}$,

$$
\begin{aligned}
x_0 &= n + (\sqrt{n^2 + 1} - n), & a_0 &= n, \\
x_1 &= \frac{1}{\sqrt{n^2+1}-n} = n + \sqrt{n^2 + 1} \\
&= 2n + (\sqrt{n^2 + 1} - n), & a_1 &= 2n, \\
x_2 &= x_1
\end{aligned}
$$

and so $\sqrt{n^2 + 1} = [n; \overline{2n}]$.
For $\sqrt{n^2 + 2}$,

$$
\begin{aligned}
x_0 &= \sqrt{n^2 + 2} = n + (\sqrt{n^2 + 2} - n), & a_0 &= n, \\
x_1 &= \frac{1}{\sqrt{n^2+2}-n} = \frac{n+\sqrt{n^2+2}}{2} \\
&= n + \frac{\sqrt{n^2+2}-n}{2}, & a_1 &= n, \\
x_2 &= \frac{2}{\sqrt{n^2+2}-n} = n + \sqrt{n^2 + 2} \\
&= 2n + (\sqrt{n^2 + 2} - n), & a_2 &= 2n, \\
x_3 &= x_1
\end{aligned}
$$

which implies that $\sqrt{n^2 + 2} = [n; \overline{n, 2n}]$.

For $\sqrt{n^2 + 2n}$,

$$
\begin{aligned}
x_0 &= \sqrt{n^2 + 2n} = n + (\sqrt{n^2 + 2n} - n), & a_0 &= n \\
x_1 &= \frac{1}{\sqrt{n^2 + 2n} - n} = \frac{\sqrt{n^2 + 2n} - n}{2n}, & a_1 &= 1 \\
x_2 &= \frac{2n}{\sqrt{n^2 + 2n} - n} = n + \sqrt{n^2 + 2n} \\
&= 2n + (\sqrt{n^2 + 2n} - n), & a_2 &= 2n \\
x_3 &= x_1,
\end{aligned}
$$

hence $\sqrt{n^2 + 2n} = [n; \overline{1, 2n}]$.

(b) Using the continued fraction representations obtained in part (a),

$$
\begin{aligned}
\sqrt{2} &= \sqrt{1^2 + 1} = [1; \overline{2}], & \sqrt{3} &= \sqrt{1^2 + 2} = [1; \overline{1, 2}] \\
\sqrt{15} &= \sqrt{3^2 + 2 \cdot 3} = [3; \overline{1, 6}], & \sqrt{37} &= \sqrt{6^2 + 1} = [6; \overline{12}].
\end{aligned}
$$

7. (a) For $e = [2; 1, 2, 1, 1, 4, 1, 1, 6, ...]$, the convergents are
   $2, 3, 8/3, 11/4, 19/7, 87/32, 106/39, 193/71, 1264/465, ...$
   Using $C_8 = 1264/465 = 2.7182796...,$

   $$|e - C_8| = 0.0000022....$$

   (b) Suppose that $e < a/b < 87/32 = C_5$. According to Theorem 15-8, we must have $b > 32$. Also $106/39 = C_6 < e$ from Theorem 15-4, so that

   $$106/39 < a/b < 87/32.$$

   These inequalities imply that $0 < 32(39a - 106b)- < b$. Now the expression $39a - 106b$ is a positive integer. If $39a - 106b = 1$, then $a = 87 + 106t$, $b = 32 + 39t$ for $t = 1, 2, ...$; thus $b \geq 71 > 39$. On the other hand, if $39a - 106b \geq 2$, then $b > 32(39a - 106b) \geq 64$, whence $b > 39$.

9. The convergents of the continued fraction

   $$x = [1; 3, 1, 5, 1, 7, 1, 9, ...]$$

   are

   $$1, 4/3, 5/4, 29/23, 34/27, 267/212, 301/239, ...$$

(a) From Theorem 15-8, the best rational approximation a/b to x, with $b < 25$, is $C_3 = 29/23$.

(b) The best rational approximation a/b, with $b < 225$, is $C_5 = 267/212$.

11. For $\pi = [3; 7, 15, 1, 292, \ldots]$, the convergents are

$$3,\ 22/7,\ 333/106,\ 355/113, \ldots.$$

Notice that

$$
\begin{aligned}
|\pi - 3| &< 1/\sqrt{5}, \quad \text{that is,} \\
&\quad 0.141592\ldots < 0.447213\ldots \\
|\pi - 22/7| &< 1/(\sqrt{5} \cdot 7^2); \quad \text{that is,} \\
&\quad 0.001264\ldots < 0.009126\ldots; \\
|\pi - 333/106| &< 1/(\sqrt{5} \cdot 106^2); \quad \text{that is,} \\
&\quad 0.0000003\ldots < 0.0000004\ldots.
\end{aligned}
$$

13. (a) Consider the irrational number

$$x = [a_0; a_1, a_2, \ldots, a_n, x_{n+1}] = \frac{x_{n+1}p_n + p_{n-1}}{x_{n+1}q_n + q_{n-1}}$$

where $C_n = p_n/q_n$ is the n'th convergent of x. Clearly

$$
\begin{aligned}
x - \frac{p_n}{q_n} &= \frac{x_{n+1}p_n + p_{n+1}}{x_{n+1}q_n + q_{n-1}} - \frac{p_n}{q_n} \\
&= \frac{(-1)(p_n q_{n-1} - q_n p_{n-1})}{(x_{n+1}q_n + q_{n-1})q_n} \\
&= \frac{(-1)^n}{(x_{n+1}q_n + q_{n-1})q_n}.
\end{aligned}
$$

Now $2a_{n+1} > a_{n+1} + (1/x_{n+2}) = x_{n+1}$, leading to

$$
\begin{aligned}
2q_{n+1} &= 2(a_{n+1}q_n + q_{n-1}) = 2a_{n+1}q_n, +2q_{n-1} \\
&> x_{n+1}q_n + q_{n-1}
\end{aligned}
$$

and so $2q_{n+1}q_n > (x_{n+1}q_n + q_{n-1})q_n$. This gives

$$\left| x - \frac{p_n}{q_n} \right| = \frac{1}{(x_{n+1}q_n + q_{n-1})q_n} > \frac{1}{2q_{n+1}q_n}.$$

(b) Since $x(x_{n+1} + q_{n+1}) = x_{n+1}p_n + p_{n-1}$, it follows that $x_{n+1}(xq_n - p_n) = -xq_{n-1} + p_{n-1}$, whence

$$|x_{n+1}q_n| \left| x - \frac{p_n}{q_n} \right| = |q_{n-1}| \left| x - \frac{p_{n-1}}{q_{n-1}} \right|.$$

But $|x_{n+1}q_n| > |1 \cdot q_{n-1}| = |q_{n-1}|$ and therefore

$$\left| x - \frac{p_n}{q_n} \right| < \left| x - \frac{p_{n-1}}{q_{n-1}} \right|.$$

15.4   Pell's Equation

1. Let x_0, y_0 be a positive solution of the equation $x^2 - dy^2 = 1$, with $d > 0$. Then $x_0^2 = 1 + dy_0^2 > dy_0^2 \geq y_0^2$, whence $x_0 > y_0$.

3. (a) Consider the equation $x^2 - 2y^2 = 1$, where $\sqrt{2} = [1; \overline{2}]$ has period 1. The convergents C_k, $0 \leq k \leq 7$, for $\sqrt{2}$ are

$$1, 3/2, 7/5, 17/12, 41/29, 99/70, 239/169, 577/408.$$

By Theorem 15-12, the positive solutions of $x^2 - 2y^2 = 1$, with $y < 250$, are obtained from C_1, C_3, C_5 and C_7; these solutions are

$$x = 3, \ y = 2; \quad x = 17, \ y = 12; \quad x = 99, \ y = 70.$$

(b) Consider the equation $x^2 - 3y^2 = 1$, where $\sqrt{3} = [1; \overline{1,2}]$ has period 2. The convergents C_k, $0 \leq k \leq 9$ for $\sqrt{3}$ are

$$1, 2/1, 5/3, 7/4, 9/11, 26/15, 71/41, 97/56, 265/153, 362/209.$$

The positive solutions of $x^2 - 3y^2 = 1$ for which $y < 250$ arise from C_1, C_3, C_5, C_7, and C_9; these are

$$x = 2, \ y = 1; \quad x = 7, \ y = 4; \quad x = 26, \ y = 15;$$
$$x = 97, \ y = 56; \quad x = 362, \ y = 209.$$

(c) For the equation $x^2 - 5y^2 = 1$, $\sqrt{5} = [2; \overline{4}]$ has period 1. The first few convergents of $\sqrt{5}$ are

$$2, \ 9/4, \ 38/17, \ 161/72, \ 682/305.$$

The solutions with $y < 250$ come from $9/4$ and $161/72$ and are

$$x = 9, \ y = 4; \quad x = 161, \ y = 72.$$

5. (a) Consider the equation $x^2 - 23y^2 = 1$, where $\sqrt{23}[4; \overline{1,3,1,8}]$ has period 4. Theorem 15-12 implies that the convergents $C_3 = 24/5$ and $C_7 = 1151/240$ furnish two solutions of the equation; namely,

$$x = 24, \ y = 5 \quad x = 1151, \ y = 240.$$

(b) For the equation $x^2 - 26y^2 = 1$, $\sqrt{26} = [5; \overline{10}]$ has period 1. Two solutions $x = 51, \ y = 10$ and $x = 5201, \ y = 1020$ are obtained from the convergents $C_1 = 51/10$ and $C_3 = 5201/1020$ of 26.

(c) Consider the equation $x^2 - 33y^2 = 1$, where $\sqrt{33} = [5; \overline{1,2,1,10}]$ has period 4. The convergents $C_3 = 23/4$ and $C_7 = 1057/184$ lead to the solutions

$$x = 23, y = 4; \quad x = 1057, y = 184.$$

7. (a) $\sqrt{13} = [3; \overline{1,1,1,6}]$ has period 5. From the lemma preceding Theorem 15-12, a particular solution of $x^2 - 13y^2 = -1$ can be obtained from the convergent C_4 of $\sqrt{13}$; that is, $C_4 = 18/5$ leads to $x = 18, \ y = 5$.

(b) Since $\sqrt{29}$ has period 5, the convergent $C_4 = 70/13$ leads to a solution $x = 70, \ y = 13$ of the equation $x^2 - 13y^2 = -1$.

(c) Since $\sqrt{41}$ has period 3, the convergent $C_2 = 32/5$ provides a solution $x = 32, \ y = 5$ of $x^2 - 41y^2 = -1$.

9. Suppose that $p|d$, where p is a prime of the form $4k + 3$. Then $x^2 - dy^2 = -1$ implies that $x^2 \equiv -1 \pmod{p}$. By Theorem 5-5 this congruence has no solution, hence the equation $x^2 - dy^2 = -1$ is not solvable.

11. Consider the pair of integers x_n, y_n defined by

$$x_n + y_n\sqrt{d} = (x_1 + y_1\sqrt{d})^n,$$

where x_1, y_1 is the fundamental solution of $x^2 - dy^2 = 1$. To see that x_n, y_n may be calculated from the relations

$$x_{n+1} = x_1x_n + dy_1y_n, \quad y_{n+1} = x_1y_n + x_ny_1 \quad (n = 1, 2, \ldots)$$

we proceed by induction. Since

$$(x_1^2 + dy_1^2) + (2x_1y_1)\sqrt{d} = (x_1 + y_1\sqrt{d})^2 = x_2 + y_2\sqrt{d},$$

it follows that $x^2 = x_1^2 + dy_1^2$, $y^2 = 2x_1 y_1$; thus, the asserted relations hold when $n = 1$. Assume that they hold when $n = k$. That is, we have $x_{k+1} = x_1 x_k + dy_1 y_k$, $y_{k+1} = x_1 y_k + x_k y_1$. Using this assumption,

$$
\begin{aligned}
(x_1 x_{k+1} \;\; + \;\; & dy_1 y_{k+1}) + (x_1 y_{k+1} + x_{k+1} y_1)\sqrt{d} \\
= \;\; & (x_{k+1} + y_{k+1}\sqrt{d})(x_1 + y_1\sqrt{d}) \\
= \;\; & [(x_1 x_k + dy_1 y_k) + (x_1 y_k + x_k y_1)\sqrt{d}](x_1 + y_1\sqrt{d}) \\
= \;\; & (x_k + y_k\sqrt{d})(x_1 + y_1\sqrt{d})^2 \\
= \;\; & (x_1 + y_1\sqrt{d})^k (x_1 + y_1\sqrt{d})^2 = (x_1 + y_1\sqrt{d})^{k+2} \\
= \;\; & x_{k+2} + y_{k+2}\sqrt{d}.
\end{aligned}
$$

The implication is that

$$
\begin{aligned}
x_{k+2} &= x_1 x_{k+1} + dy_1 y_{k+1}, \\
y_{k+2} &= x_1 y_{k+1} + x_{k+1} y_1;
\end{aligned}
$$

so the assertion is true for $k + 1$ whenever it is true for k, completing the induction step. From these recursion formulas for the pair of integers x_n, y_n, another set may be obtained. Namely, for $n \geq 2$,

$$
\begin{aligned}
x_{n+1} &= x_1 x_n + dy_1 y_n = 2x_1 x_n - (x_1 x_n - dy_1 y_n) \\
&= 2x_1 x_n - [x_1(x_1 x_{n-1} + dy_1 y_{n-1}) - dy_1(x_1 y_{n-1} + x_{n-1} y_1)] \\
&= 2x_1 x_n - x_{n-1}(x_1^2 - dy_1^2) = 2x_1 x_n - x_{n-1};
\end{aligned}
$$

and also

$$
\begin{aligned}
y_{n+1} &= x_1 y_n + x_n y_1 = 2x_1 y_n - (x_1 y_n - x_n y_1) \\
&= 2x_1 y_n - [x_1(x_1 y_{n-1} + x_{n-1} y_n) - (x_1 x_{n-1} + dy_1 y_{n-1})y_1] \\
&= 2x_1 y_n - y_{n-1}(x_1^2 - dy_1^2) = 2x_1 y_n - y_{n-1}.
\end{aligned}
$$

13. (a) Assume that the equation $x^2 - dy^2 = c$ has a solution u, v and that r, s satisfies $x^2 - dy^2 = 1$; then $u^2 - dv^2 = c$ and $r^2 - ds^2 = 1$. This implies that

$$
\begin{aligned}
(ur \;\; + \;\; dvs)^2 &- d(us + vr)^2 \\
&= u^2(r^2 - ds^2) - dv^2(r^2 - ds^2) \\
&= (u^2 - dv^2)(r_2 - ds^2) = c \cdot 1 = c,
\end{aligned}
$$

whence $x = ur + dvs$, $y = us + vr$ is another solution of $x^2 - dy^2 = c$. By Theorem 15-12, there are infinitely many solutions to $x^2 - dy^2 = 1$ and, in turn, to $x^2 - dy^2 = c$.

(b) The equation $x^2 - 7y^2 = 4$ is satisfied by $x = 16$, $y = 6$. Since $x = 8$, $y = 3$ and $x = 127$, $y = 48$ are both solutions of $x^2 - 7y^2 = 1$, part (a) indicates that two other solutions of $x^2 - 7y^2 = 4$ are

$$x = 16 \cdot 8 + 7 \cdot 6 \cdot 3 = 254, \qquad y = 1 \cdot 3 + 6 \cdot 8 = 96 \ \text{ and}$$
$$x = 16 \cdot 127 + 7 \cdot 6 \cdot 48 = 4048, \quad y = 16 \cdot 48 + 6 \cdot 127 = 1530.$$

(c) The equation $x^2 - 35y^2 = 9$ has $x = 18$, $y = 3$ as a solution, while $x = 6$, $y = 1$ and $x = 71$, $y = 12$ provide two solutions to $x^2 - 35y^2 = 1$. Those two other solutions of $x^2 - 35y^2 = 9$ are given by

$$x = 18 \cdot 6 + 35 \cdot 3 \cdot 1 = 213, \qquad y = 18 \cdot 1 + 6 \cdot 3 = 36 \ \text{ and}$$
$$y = 18 \cdot 71 + 35 \cdot 3 \cdot 12 = 2538, \quad y = 18 \cdot 12 + 3 \cdot 71 = 429.$$

15. Consider the roots $\alpha = 1 + \sqrt{2}$ and $\beta = 1 - \sqrt{2}$ of the quadratic equation $x^2 - 2x - 1 = 0$. Then $\alpha^2 = 2\alpha + 1$ and $\beta^2 = 2\beta + 1$. Multiply the first of these relations by α^n and the second by β^n to obtain

$$\alpha^{n+2} = 2\alpha^{n+1} + \alpha^n \ \text{ and } \ \beta^{n+2} = 2\beta^{n+1} + \beta^n.$$

Subtracting the second equation from the first and dividing by $\alpha - \beta$ leads to
$$\frac{\alpha^{n+2} - \beta^{n+2}}{\alpha - \beta} = 2\frac{\alpha^{n+1} - \beta^{n+1}}{\alpha - \beta} + \frac{\alpha^n - \beta^n}{\alpha - \beta}.$$

If we set $H_n = (\alpha^n - \beta^n)/(\alpha - \beta)$, this last equation becomes $H_{n+2} = 2H_{n+1} + H_n$. Also, $H_1 = 1$ and $H_2 = (\alpha^2 - \beta^2)/(\alpha - \beta) = \alpha + \beta = 2$. Thus, the sequence $\{H_n\}$ is the sequence of Pell numbers p_n, which implies that $p_n = (\alpha^n - \beta^n)/(\alpha - \beta) = (\alpha^n - \beta^n)/2\sqrt{2}$. A similar argument gives $q_n = (\alpha_n + \beta_n)/2$.

Chapter 16

Some Twentieth-Century Developments

16.2 Primality Testing and Factorization

1. (a) To factor 299, take $x_0 = 2$ and $f(x) = x^2 + 1$. The successive iterates $x_{k+1} = f(x_k)$ are, modulo 299,

$$x_0 \equiv 2, x_1 \equiv 5, x_2 \equiv 26, x_3 \equiv 79, x_4 \equiv 262, x_5 \equiv 174, \ldots.$$

We obtain

$$\gcd(x_5 - x_1, 299) = \gcd(169, 299) = 13,$$

so that 13 is a factor of 299; a complete factorization is $299 = 13 \cdot 23$.

(b) Let $x_0 = 2$, $f(x) = x^2 + 1$ and $x_{k+1} = f(x_k)$. Working modulo 1003, the early values of the x_k are

$$x_0 \equiv 2, x_1 \equiv 5, x_2 \equiv 26, x_3 \equiv 677, x_4 \equiv 962, x_5 \equiv 324,$$

$$x_6 \equiv 665, x_7 \equiv 906, x_8 \equiv 383, x_9 \equiv 252, \ldots$$

Since $\gcd(x_9 - x_3, 1003) = \gcd(-425, 1003) = 17$, it follows that 17 divides 1003; indeed, $1003 = 17 \cdot 59$.

(c) Take $x_0 = 2$ and $f(x) = x^2 + 1$. For $x_{k+1} \equiv f(x_k) \pmod{8051}$, the required sequence is

$$x_0 \equiv 2, x_1 \equiv 5, x_2 \equiv 26, x_3 \equiv 677, x_4 \equiv 7474,$$

$$x_5 \equiv 2839, x_6 \equiv 871, \ldots$$

Here, $\gcd(x_6 - x_3, 8051) = \gcd(194, 8051) = 97$, which leads to the factorization $8051 = 83 \cdot 97$.

3. (a) To factor 1711, compute $2^{7!}$ (mod 1711). The sequence of computations, modulo 1711, is

$$2^2 \equiv 4, 4^3 \equiv 64, 64^4 \equiv 861, 861^5 \equiv 488,$$

$$488^6 \equiv 1038, 1038^7 \equiv 1074.$$

Since $\gcd(1073, 1711) = 29$, it follows that 29 is a divisor of 1711; in fact, $1711 = 29 \cdot 59$.

(b) It suffices to evaluate $2^{6!}$ (mod 4847) in order to factor 4847. Working modulo 4847, the necessary calculations are

$$2^2 \equiv 4, 4^3 \equiv 64, 64^4 \equiv 1749, 1749^5 \equiv 322, 322^6 \equiv 2850.$$

Now $\gcd(2849, 4847) = 37$, so that $4847 = 37 \cdot 131$.

(c) We factor 9943 by obtaining $2^{5!}$ (mod 9943). The computations proceed, modulo 9943, in the following order:

$$2^2 \equiv 4, 4^3 \equiv 64, 64^4 \equiv 3375, 3375^5 \equiv 2685,$$

with $\gcd(2684, 9943) = 61$. This leads to the factorization $9943 = 61 \cdot 163$.

5. We do not have to go far in the continued fraction expansion $\sqrt{7134} = [84; 2, 6, 31, \ldots]$ to discover that $p_1^2 \equiv (-1)^2 t_2 = 5^2$ (mod 7134) or, expressed differently,

$$169^2 \equiv 5^2 \pmod{7134}.$$

Since

$$\gcd(169 + 5, 7134) = \gcd(174, 6 \cdot 1189) = 29$$
$$\gcd(169 - 5, 7134) = \gcd(164, 6 \cdot 1189) = 41,$$

1189 has divisors 29 and 41. In fact, $1189 = 29 \cdot 41$.

7. (a) Given 907, we have $907 - 1 = 9016 = 2 \cdot 3 \cdot 151$. Since

$$2^6 \equiv 64 \pmod{907}, 2^{302} \equiv 384 \pmod{907}, 2^{453} \equiv -1 \pmod{907},$$

Lucas's Theorem implies that 907 is prime.

(b) Here, $1301 - 1 = 1300 = 2^2 \cdot 5^2 \cdot 13$. Letting $a = 2$, we calculate

$$2^{650} \equiv -1 \pmod{1301}, \ 2^{260} \equiv 163 \pmod{1301},$$

$$2^{100} \equiv 78 \pmod{1301}$$

from which it follows that 1301 is prime.

(c) In this case, $1709 - 1 = 1708 = 2^2 \cdot 7 \cdot 61$. If 3 is the base, then

$$3^{854} \equiv -1 \pmod{1709}, \ 3^{244} \equiv 880 \pmod{1709},$$

$$3^{28} \equiv 753 \pmod{1709}$$

This establishes the primality of 1709.

9. Let $n = k \cdot 2^m + 1$, which gives $n - 1 = k \cdot 2^m$. From the inequality $1 \leq k < 2^m$, we find that $n < 2^{2m} + 1$ and so $n \leq 2^{2m}$; thus $\sqrt{n} \leq 2^m$. If there exists some integer a satisfying $a^{(n-1)/2} \equiv -1 \pmod{n}$, then

$$\gcd(a^{(n-1)/2} - 1, n) = \gcd(-2, n) = 1.$$

Hence the hypothesis of Pocklington's Theorem would hold, making n prime.

11. (a) The integer $2047 = 2^{11} - 1 = 23 \cdot 89$ is composite. Also $2046 = 2 \cdot 1023$. Here, $3^{1023} \equiv 1565 \not\equiv 1 \pmod{2047}$, so that 2047 fails the Miller-Rabin test to the base 3.

(b) For $25 = 5^2$, we have $24 = 2^3 \cdot 3$. Since $7^{2 \cdot 3} = 7^6 \equiv -1 \pmod{25}$, the integer 25 passses the Miller-Rabin test to the base 7.

(c) For 65, $64 = 2^6$. It follows that $8^2 \equiv -1 \pmod{65}$ and $18^2 \equiv -1 \pmod{65}$, which means that 65 passes the Miller-Rabin test to both the bases 8 and 18.

(d) It is already known that $341 = 11 \cdot 31$ is a pseudoprime to the base 2. Now $340 = 2^2 \cdot 85$. Since $2^{85} \equiv 32 \not\equiv 1 \pmod{341}$ and $2^{170} \equiv 1 \not\equiv -1 \pmod{341}$, the integer 341 fails the Miller-Rabin test to the base 2.

13. According to Problem 4(b) of Section 11.3, a composite Fermat number F_n is a pseudoprime (base 2). Consequently, F_n is a strong pseudoprime by Problem 9 of this section.

16.3   An Application to Factoring: Remote Coin Flipping

1. The congruence $x^2 \equiv 12 \pmod{85}$ will have a solution if $x^2 \equiv 12 \equiv 2 \pmod 5$ and $x^2 \equiv 12 \pmod{17}$ are both solvable. But the Legendre symbols $(2/5) = -1$ and $(12/17) = (3/17) = -1$ indicate that neither of these congruences has a solution.

3. Since $73^2 = 5329 \equiv 338 \pmod{713}$, where $713 = 23 \cdot 31$, Bob sends Alice the value 338. She then solves the two congruences $x^2 \equiv 338 \equiv 16 \pmod{23}$ and $x^2 \equiv 338 \equiv 28 \pmod{23}$ for $x \equiv \pm 4 \pmod{23}$ and $x \equiv \pm 11 \pmod{31}$, respectively. Combining these numbers in four ways, Alice forms four pairs of linear congruences whose solutions — via the Chinese Remainder Theorem — are $x \equiv 42, 73, 640$ and $671 \pmod{713}$; they satisfy $x^2 \equiv 338 \pmod{713}$. If Alice picks either 73 or $640 \equiv -73 \pmod{713}$ to forward to Bob, she wins the coin toss. But if Bob receives 42 or 671, he is able to factor 713:

$$\gcd(73 + 42, 713) = \gcd(115, 713) = 23$$
$$\gcd(73 + 671, 713) = \gcd(774, 713) = 23$$

5. Alice calculates both $110 \equiv 2^{63} \pmod{173}$ and $143 \equiv 3^{63} \pmod{173}$. She sends one of these values, say 143, to Bob. If Bob guesses that 143 was obtained from the primitive root 3, then he wins; otherwise, Bob loses the coin toss.

Miscellaneous Problems

1. (a) If $n = 1$, then $1 \cdot 2 \cdot 3 = \frac{1 \cdot 2 \cdot 3 \cdot 4}{4}$, so the assertion holds when $n = 1$. Suppose that for some k,

$$1 \cdot 2 \cdot 3 + 2 \cdot 3 \cdot 4 + \cdots + k(k+1)(k+2) = \frac{k(k+1)(k+2)(k+3)}{4}.$$

Then

$$
\begin{aligned}
1 \cdot 2 \cdot 3 \quad & + \quad 2 \cdot 3 \cdot 4 + \cdots + k(k+1)(k+2) + (k+1)(k+2)(k+3) \\
& = \frac{k(k+1)(k+2)(k+4)}{4} + (k+1)(k+2)(k+3) \\
& = (k+1)(k+2)(k+3)\left[\frac{k}{4} + 1\right] \\
& = \frac{(k+1)(k+2)(k+3)(k+4)}{4}
\end{aligned}
$$

This shows that if the assertion is correct for k, then it is also correct for $k+1$, thereby completing the induction.

(b) If $n = 1$, then $\frac{1}{1 \cdot 5} = \frac{1}{4+1}$, so the assertion is true for $n = 1$. Assume that for some k, we have

$$\frac{1}{1 \cdot 5} + \frac{1}{5 \cdot 9} + \cdots + \frac{1}{(4k-3)(4k+1)} = \frac{k}{4k+1}.$$

Then

$$
\begin{aligned}
\frac{1}{1 \cdot 5} + \frac{1}{5 \cdot 9} + \quad \cdots \quad & + \frac{1}{(4k-3)(4k+1)} + \frac{1}{(4k+1)(4k+5)} \\
& = \frac{k}{4k+1} + \frac{1}{(4k+1)(4k+5)} \\
& = \frac{1}{4k+1}\left[k + \frac{1}{4k+5}\right]
\end{aligned}
$$

131

$$= \frac{1}{4k+1} \left[\frac{4k^2 + 5k + 1}{4k+5} \right]$$

$$= \frac{1}{4k+1} \left[\frac{(4k+1)(k+1)}{4k+5} \right] = \frac{k+1}{4k+5}.$$

This completes the induction and proves the assertion.

(c) When $n = 1$, the assertion reduces to $1 \geq \sqrt{1}$, so is correct in this case. For some k supose that

$$1 + \frac{1}{\sqrt{2}} + \cdots + \frac{1}{\sqrt{k}} \geq \sqrt{k}.$$

Then

$$1 + \frac{1}{\sqrt{2}} + \cdots + \frac{1}{\sqrt{k}} + \frac{1}{\sqrt{k+1}} = \sqrt{k} + \frac{1}{\sqrt{k+1}}$$

$$= \frac{\sqrt{k}(\sqrt{k+1}) + 1}{\sqrt{k+1}}$$

$$\geq \frac{\sqrt{k}(\sqrt{k}) + 1}{\sqrt{k+1}} = \frac{k+1}{\sqrt{k+1}} = \sqrt{k+1}.$$

Hence the assertion holds also for $k + 1$. This completes the induction.

3. (a) If $n = 1$, Then $7|(2^4 + 4^4 + 1)$ or $7|243$, so the assertion holds in this case. Assume that for some k, $7|2^{3k+1} + 4^{3k+1} + 1$. Then

$$2^{3(k+1)+1} + 4^{3(k+1)+1} + 1 = 8 \cdot 2^{3k+1} + 64 \cdot 4^{3k+1} + 1$$

$$= 64 \left[2^{3k+1} + 4^{3k+1} + 1 \right] - \left[56 \cdot 2^{3k+1} + 63 \right],$$

where each summand is divisible by 7; hence the assertion is correct for $k + 1$.

(b) When $n = 1$, $133|(11^3 + 12^3)$ or $133|3059$, making the assertion correct for $n = 1$. Suppose we know that $133|11^{k+2} + 12^{2k+1}$ for some k. Then

$$11^{(k+1)+2} + 12^{2(k+1)+1} = 11 \cdot 11^{k+2} + 144 \cdot 12^{2k+1}$$

$$= 144 \left[11^{k+2} + 12^{2k+1} \right] - 133 \cdot 11^{k+2},$$

with each summand divisible by 133. Thus the truth of the assertion for k implies its truth for $k+1$, completing the induction.

(c) The assertion is correct for $n = 1$, since $11|(3^5 + 4^7 + 5^6)$ or $11|32252$. Assume for some k that $11|(3^{5k}+4^{5k+2}+5^{5k+1})$. Then

$$3^{5(k+1)} + 4^{5(k+1)+2} + 5^{5(k+1)+1}$$
$$= 3^5 \cdot 3^{5k} + 4^5 \cdot 4^{5k+2} + 5^5 \cdot 5^{5k+1}$$
$$= 3125\left[3^{5k} + 4^{5k+2} + 5^{5k+1}\right] - \left[2882 \cdot 3^{5k} + 2101 \cdot 4^{5k+2}\right]$$

with each summand divisible by 11, so the assertion holds for $k + 1$.

5. Let us use congruence theory to show that $3|8 \cdot 2^{2^n} + 1$. Working modulo 3, it is clear that

$$8 \cdot 2^{2^n} + 1 = 8 \cdot (2^2)^{2^{n-1}} + 1 \equiv 8 \cdot 1 + 1 \equiv 0 \pmod{3}.$$

7. The integer n is of the form $3k$, $3k + 1$ or $3k + 2$. In the latter two cases

$$(3k + 1)^2 + 2 = 3(3k^2 + 2k + 1)$$

and

$$(3k + 2)^2 = 3(3k^2 + 4k + 2),$$

neither of which is prime. Thus $n = 3k$ for some k.

9. (a) By Problem 8(b) of Section 4.2, $n^3 \equiv 0, 1$ or $6 \pmod{7}$ for any integer n. In order that $a^3 + b^3 + c^3 \equiv 0 \pmod{7}$, at least one of a, b or c is congruent to 0 modulo 7.

(b) Observe that $(n-1)^3 + n^3 + (n+1)^3 = 3n(n^2+2)$. Now n is of the form $3k, 3k + 1$ or $3k + 2$; in each case $3n(n^2 + 2) \equiv 0 \pmod{9}$.

11. (a) Since $\gcd(301, 77) = 7$, th equation $301x + 77y = 2000 + n$ will have a solution if 7 divides $2000 + n$. The smallest positive n for which this holds is 9.

(b) One solution of $5x + 7y = n$ is $x = 3n$, $y = -2n$, so the general solution is $x = 3n + 7t$, $y = -2n - 5t$. In order that $x > 0, y > 0$, t must satisfy $-3n/7 < t < -2n/5$; thus, t lies in an interval of length $n/35$. When $n = 140$, there are three values of t giving positive solutions, namely $t = -59, -58$ and -57.

13. If $3|242628x91715131$, then $52 + x \equiv 0 \pmod{3}$ or $x \equiv 2 \pmod{3}$. Thus $x = 2, 5$ or 8.

15. The problem is equivalent to solving the system of congruences

$$
\begin{aligned}
x &\equiv 3 \pmod{5} \\
x &\equiv 3 \pmod{7} \\
x &\equiv 3 \pmod{9}.
\end{aligned}
$$

This system has solution $x = 3 + 315t$, with 318 inches being the meaningful solution.

17. Suppose that $R_n = p = a^2 + b^2$. Since $R_n \equiv 3 \pmod{4}$, Problem 12 of Section 5.4 implies that $R_n^2 | R_n$, which is impossible.

19. The numbers being squared are of the form $3R_n + 1$; and

$$
\begin{aligned}
(3R_n + 1)^2 = 9R_n^2 + 6R_n + 1 &= (10^n - 1)R_n + 6R_n + 1 \\
&= 10^n R_n + 5R_n + 1.
\end{aligned}
$$

21. The formula $\sum_{d|n} |\mu(d)| = 2^{\omega(n)}$ occurs in Problem 5 of Section 6.2. If the Möbius Inversion Theorem is applied, the result becomes

$$
\sum_{d|n} \mu(d) 2^{\omega(n/d)} = |\mu(n)|.
$$

23. If $\tau(n)$ is divisible by an odd prime, then n has a square factor; hence $\mu(n) = 0$.

25. If n is replaced by $k+4$, the equation $(n-1)^3 + n^3 + (n+1)^3 = (n+2)^3$ becomes $k(2k^2 + 18k + 42) = 0$. The only integer solution is $k = 0$, so that $n = 4$.

27. Consider an even perfect number $n = 2^{k-1}(2^k - 1)$, where $2^k - 1$ is a prime. If n is square-free, then

$$
2^{k-1}(2^k - 1) = pq
$$

for distinct primes $p < q$. Since $2^{k-1} = p$, it follows that $p = 2$ and $k - 1 = 1$; that is, $k = 2$. Then $2^k - 1 = 3$, making $n = 2 \cdot 3 = 6$.

29. The congruence $2^n L_n \equiv 2 \pmod{10}$ is obtained by induction on n. For the induction step, assume the congruence holds when n is one of the integers $1, 2, \ldots, k$, where $k \geq 2$. Then

$$
\begin{aligned}
2^{k+1} L_{k+1} = 2^{k+1}(L_k + L_{k-1}) &= 2 \cdot 2^k L_k + 4 \cdot 2^{k-1} L_{k-1} \\
&\equiv 2 \cdot 2 + 4 \cdot 2 \equiv 12 \equiv 2 \pmod{10}.
\end{aligned}
$$

Thus the congruence is also true for $n = k + 1$.

31. When the identity $u_{n+k} = u_{n-1}u_k + u_n u_{k+1}$ is used to substitute for the first four terms in the expression

$$u_{n+11} + u_{n+7} + 8u_{n+5} + u_{n+3} + 2u_n$$

the result is

$$u_{n-1}(u_{11} + u_7 + 8u_5 + u_3) \quad + \quad u_n(u_{12} + u_8 + 8u_6 + u_4 + 2)$$
$$= \quad 144u_{n-1} + 234u_n$$
$$= \quad 18(8u_{n-1} + 13u_n).$$

33. Suppose that $n = 10k + 5$ for some k. By Fermat's theorem, $a^{10k+5} \equiv a^5 \pmod{11}$ whenever $\gcd(a, 11) = 1$. Thus,

$$12^n + 9^n + 8^n + 6^n \quad \equiv \quad 12^5 + 9^5 + 8^5 + 6^5$$
$$\equiv \quad 1^5 + (-2)^5 + (-3)^5 + (-5)^5$$
$$\equiv \quad 1 + 1 + (-1) + (-1) \equiv 0 \pmod{11}$$

35. For any integer k, $k^3 \equiv 0$, 1 or $-1 \pmod 9$. Considering the possible combination of values of a, b, c, it follows that $a^3 + b^3 + c^3 \not\equiv \pm 4 \pmod 9$.

37. The equation $(9999)n = R_{4k}$, equivalently $(10^4 - 1)n = (10^{4k} - 1)/9$, implies that

$$n = (1/9)(10^{4(k-1)} + 10^{4(k-2)} + \cdots 10^4 + 1).$$

The right-hand side will be an integer provided that 9 divides the expression $10^{4(k-1)} + \cdots + 10^4 + 1$. This happens when 9 divides k, for there are k terms in the expression (hence k 1's). A choice for n is

$$n = (1/9)(10^{32} + 10^{28} + \cdots + 10^4 + 1).$$

39. For $p \geq 5$, assume that $p^\# + 1 = (2 \cdot 3 \cdot 5 \cdots p) + 1 = a^2$ for some integer a. Since $a^2 \equiv 1 \pmod 6$, we have $a \equiv \pm 1 \pmod 6$. Letting $a = 6k \pm 1$ for some k implies that the product $5 \cdot 7 \cdots p$ is even, an obvious impossibility.

41. For $n \geq 2$, $3 \cdot 10^n - 1 \equiv -1 \pmod{12}$. Thus

$$299 + 2999 + \cdots + 29999999999999 \equiv 12(-1) \equiv 0 \pmod{12}.$$

43. Notice that

$$(2k + 1)2^n + 1 = 2^{n+1}k + (2^n + 1) \text{ for } k = 0, 1, 2, \ldots.$$

Since $\gcd(2^{n+1}, 2n + 1) = 1$, Dirichlet's theorem guarantees that this progression contains infinitely many primes.

45. If $p = 2$, $2^2 + 2^2 = 8$ is not prime; but, if $p = 3$, $2^2 + 2^3 = 17$ is prime. Suppose that $p > 3$. Then p is of the form $6k + 1$ or $6k + 5$. In either case,

$$p^2 + 2^p \equiv 1 + 2 \equiv 0 \pmod{3}$$

and 3 divides $p^2 + 2^p$. Thus, $p^2 + 2^p$ is prime only when $p = 3$.

47. Clearly $n!$ is not a square when $n = 2$ or 3, so it may be assumed that $n > 3$. If n is even, then Bertrand's Conjecture implies that there exists a prime p such that $n/2 < p < n$. Similarly, when n is odd there exists a prime p such that $(n + 1)/2 < p < n + 1$. In either case, $n < 2p$; whence p divides $n!$, but p^2 fails to divide $n!$. Thus, $n!$ cannot be a square.

49. The integers $p_1 p_2 \cdots p_n + 1$ and $p_1 p_2 \cdots p_n - 1$ are divisible by primes q and q', respectively, different from $p_1, p_2, \ldots, p_n$. If $q = q'$, then $q | (p_1 p_2 \cdots p_n + 1) - (p_1 p_2 \cdots p_n - 1)$ or $q = 2$' it follows that $q = 2 = p_2$, which is impossible.